The ATLAS of ATLASES

A DESCRIPTION OF THE SEA COASTES OF Galicia beginning from Cap de finisterre Vnto Camino, Done According Vnto the true Situa= tion & appearing Thereof

·1588·

CIVITA
TES OR
BIS TER
RARVM
24 Augusti 1576.

CONSOCIAT. HVMANI GEN. ORIGO.

ARCHITECT. RVDIM.

DOMICIL. TYROCIN.

LIBER PRIMVS.

The ATLAS of ATLASES

The Map Maker's Vision of the World

Atlases from The Cadbury Collection,
Birmingham Central Library

Phillip Allen

Harry N. Abrams, Inc.
Publishers, New York

Library of Congress Cataloging-in-Publication Data
Allen, Phillip.
    The atlas of atlases: the map maker's vision of the world / by
Phillip Allen.
    p.   cm.
    ISBN 0-8109-3918-5
    1. Maps—Facsimiles.   2. Cartography—History.   I. Title.
G1025.A36   1992   <G&M>
912—dc20   92-9605
                                                                CIP
                                                                MAP

Published in 1992 by Harry N. Abrams, Incorporated,
New York
A Times Mirror Company

The Atlas of Atlases is a Marshall Edition

Editor   Gwen Rigby
Assistant editor   Anthony Livesey
Art director   David Fordham

Printed and bound in Germany

CONTENTS

FOREWORD

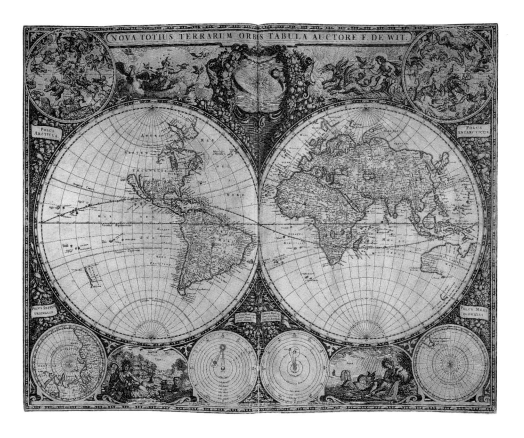

Frederick de Wit's fine map of the world was produced in 1660 and first printed in the atlas of his contemporary, the map maker Hendrick Doncker.

The map was used again in de Wit's own atlas of 1671; such cooperation between map makers and reuse of plates was typical of the period.

I T GIVES ME PLEASURE to introduce the *Atlas of Atlases*, which is based on the fine collection of printed maps and atlases preserved in Birmingham Central Library. The Central Library was founded in 1862 but the collections really date from 1879, for in January of that year the building suffered a disastrous fire in which most of the stock was destroyed.

The citizens of Birmingham had developed a great affection for the library service since its foundation in 1861, and some of the more wealthy residents were quick to make donations of books to replace those that had been lost. It is worth recording that other than local donors also contributed toward the restocking of the library, and that the list was headed by Queen Victoria, who sent several appropriate volumes.

At the time when the rebuilt Central Library was opened in 1882, it contained more than 50,000 volumes, but by 1900 the stock had increased to 150,000. Much of this growth may be attributed to a number of generous gifts, such as that devoted to the writings of Lord Byron given by Sir Richard Tangye, a noted local industrialist who arrived in the town about 1858. It is interesting to note that this same person was instrumental in securing the industrial archive of the company founded in Birmingham by Matthew Boulton and James Watt which is crucial to our understanding of the Industrial Revolution.

The Library's special collections are numerous, but no mention of them would be complete without reference to the Shakespeare Collection, which actually came into existence before the Library was founded. A proposal to establish a Shakespeare Memorial Library in Birmingham was made as early as 1858, and it was publicly endorsed in 1861, when George Dawson, one of the city's great advocates of public library provision, wrote to a local newspaper.

From the time of its foundation, the Central Library has sought to acquire books which were beyond the reach of all except the most wealthy, and this policy has been pursued to the present day. In the last two decades of the nineteenth and the first four decades of the present century, public funding and private patronage ensured that the collections grew dramatically, and the commitment and generosity of the many individuals who gave books for the enjoyment of future generations of citizens has been of vital importance to the Library.

The present volume has been published to celebrate and draw attention to the atlases that were given to the Library by William Aldington Cadbury, Lord Mayor of Birmingham (1919–21) and a former chairman of Cadbury Brothers, the famous cocoa company established at Bourneville, then on the periphery of the industrial city.

It was here that the splendid chocolate factory was built in the midst of a Model Village which housed the workforce. The Cadbury family were employers possessed of high ideals who realized the importance of healthy and contented employees, and it was for this reason that they constructed their factory some distance from the then overcrowded and grimy centre of the city.

William Cadbury was born in 1867 and served on the City Council, from 1911 as a member and later as an alderman, until his retirement in 1943. Throughout the 1920s and 1930s he acted as chairman of the Public Libraries Committee and ably assisted the bibliographically-minded City Librarian, Herbert Maurice Cashmore, in building up the Central Library's stock of early printed books.

William Cadbury's interest in geography and atlases may well have been stimulated by the fact that the family company obtained the cocoa beans from Nigeria, and he often visited the plantations there as well as travelling to other parts of the world to promote the Cadbury products. As a member of the wealthy and celebrated Quaker family, he was well able to indulge his passion for maps and atlases. He bought almost all the early, and also many of the later, examples that are now in the Birmingham collection. It is pleasing to record that he was a careful and selective buyer, acquiring only fine copies, almost always in good contemporary bindings.

The Library has augmented the Cadbury collection by continuing to acquire fine atlases of all periods, right up to the present day. The result is an accumulation unmatched in the public library sector in its scope and quality.

The text of this volume has largely been written by Phillip Allen, Special Collections Officer at the Central Library, but he has been assisted by all the staff who have responsibility for map collections, particularly by Pamela Williams, whose work embraces the care of the Early and Fine Printing Collection.

Finally, I should like to say that this account of the Birmingham atlas collection is not intended as a scholarly essay on cartography or geography, but rather as a visual celebration of the books which the Library is so fortunate to possess.

Patricia M. Coleman

Director of Birmingham Library Services

N HOC OPERE HAEC CONTI NENTVR

GEographiæ Cl.Ptolemæi a plurimis uiris utriufq; linguæ doctiff. emédata:& cú archetypo græco ab ipfis collata.

SChemata cú demonftrationibus fuis correcta a Marco Beneuentano Monacho cæleftino:& Ioanne Cotta Veronenfi uiris Mathematicis confultiffimis.

FIgura de proiectione fphęræ in plano quæ in libro octauo defidera batur ab ipfis nódum inftaurata fed fere ad inuenta eius, n.ueftigia in nullo etiam græco codice extabant.

MAxima quantitas dieg ciuitatú:& diftantiæ locog ab Alexádria Aegypti cuiufcp ciuitatis:quæ in alijs codicibus nó erant.

PLanifphærium Cl.Ptolemęi nouiter recognitú & diligentiff. emen, datum a Marco Beneuentano Monacho cęleftino.

NOua orbis defcriptio ac noua Oceani nauigatio qua Lifbona ad Indicú peruenitur pelagus Marco Beneuentano monacho cæle, ftino ædita.

NOua & uniuerfalior Orbis cogniti tabula Ioã.Ruyfch Germano elaborata.

S·Ex Tabulæ nouiter confectæ uidelicet Liuoniæ:Hyfpaniæ:Galliæ: Germaniæ:Italiæ:& Iudeæ.

CAVTVM EST EDICTO IVLII.II.PONT.MAX.
NE Q VIS IMPRIMERE AVT IMPRIMI
FACERE AVDEAT HOC IPSVM OPVS
PENA EXCOMMVNICATIONIS LATAE SENTENTIAE
HIS Q VI CONTRA MANDATVM IVSSVMQ VE
CONARI AVDEBVNT.

ANNO VIRGINEI PARTVS

M D VIII.

ROME

CHAPTER ONE

The CLASSICAL and MEDIEVAL TRADITION

MAPS IN A RUDIMENTARY FORM have existed since the earliest times, and it is recorded that the ancient Egyptians used maps to aid their tax collectors. It was, however, the Greek philosophers, in particular Pythagoras of Samos, who evolved the theory that the universe was spherical, with the earth at the centre, and it was the Greeks also who developed the science of map making.

At the time of his death in 323 BC, the empire of Alexander the Great extended into India, and it is recorded that map makers accompanied him on his campaigns, but their maps have not survived. However, the Greek historian, philosopher and general, Arrian, preserved much that was written. His major book, *Anabasis*, which relies chiefly on the published writings of two of Alexander's generals, Ptolemy I and Aristobulus, is the prime source of information about Alexander's life and times, and some of this information was incorporated into the later geographical writings of Claudius Ptolemy.

From the extensive writings of Strabo (born *c*63 BC) we know that in the third century BC Erastosthenes of Cyrene measured the circumference of the earth to within some 200mls/ 320km of the accepted modern measurement. Strabo was a Greek philosopher, geographer and historian who had studied in Asia Minor, Greece, Rome and Alexandria and had travelled extensively throughout much of Europe, north Africa and western Asia. Most of his writings are lost (he wrote 47

The contents page to the Ptolemy of 1508, *opposite*, with its decorated initial "I", details the features of the atlas with pride. Johan Ruysch's "new map of the whole known world" is specially mentioned, as are six new maps of Livonia, Spain, France, Germany, Italy and Judea.

Decorated dropped capital letters, such as the one above, were a recurring theme in the 1482 Ptolemy.

books of historical sketches) but his *Geographia*, based on his own observations during his travels, contains historical material and descriptions of places he had visited and people he had met.

The whole of the *Geographia* still exists, save for part of the seventh book; it is divided into 17 books, two describing the scope of geography, eight on Europe, six on Asia and one on Africa. Collectively they comprise a rich source of knowledge of the ancient world and its geographical awareness, although the work is uneven in that Strabo assumed that Homer had accurate knowledge of the places and people he wrote about in his epics, the Iliad and the Odyssey, while largely ignoring the historian Herodotus's first-hand information.

Erastosthenes (*c*275–*c*195 BC) studied in Athens and was renowned for his versatility, for not only did he write poetry but also works on literature, the theatre, astronomy, geography, mathematics and philosophy. He also composed a map of the known world, but he is most famous as an astronomer and for calculating the circumference of the earth, work he carried out while he was librarian at the famous Alexandrian library.

In about 190 BC Hipparchus, a Greek astronomer who made his observations mainly on the island of Rhodes, developed the work of Erastosthenes, establishing the measurement of latitude and longitude and dividing the world into 360 degrees. He also divided the world as he knew it into climatic zones. A record of his researches is preserved in Ptolemy's *Almagest*. Among much

else, he is credited with the discovery of the precession of the equinoxes and the inequalities of the motion of the moon. Hipparchus made the first known detailed chart of the heavens, fixing the position of at least 850 stars, and is thought to have been the first to make systematic use of trigonometry.

The Romans certainly used maps widely, and the emperor Augustus (r27 BC–AD 14) had a map of the empire displayed in the Via Flaminia to demonstrate the extent of Rome's power. This map did not attempt to go beyond the boundaries of the Roman world, but the writer Pliny the Elder (AD 23/4–79), a Roman naturalist, soldier and friend of Emperor Vespasian, in his one surviving work—a 37-volume encyclopedia of natural science—enlivened his text with stories about fantastic creatures who were denizens of distant lands. These included the one-legged sciopodes who lay on their backs and used their single foot to shade themselves from the sun; many such strange beings were still being faithfully illustrated in the printed atlases of the sixteenth century. Pliny's industry was awesome, and his knowledge of sources without rival, but his information was almost entirely secondhand.

The second-century Greco-Egyptian geographer Claudius Ptolemy of Alexandria continued the work of his older, and inaccurate, contemporary Marinus of Tyre, who wished to produce a map by calculating the latitude and longitude of all known places. In light of the importance which his *Geographia* assumed in later centuries, it was unfortunate that the more accurate calculations of Erastosthenes were rejected by Ptolemy.

Six of the eight books of the *Geographia* contain authoritative tables of latitude and longitude and brief descriptions of a vast number of places, but some of the coordinates were incorrect and an error of 90 degrees resulted in Scotland appearing east of the rest of the British Isles. Ptolemy also discarded the correct theory

that Africa was surrounded by water and joined its east coast to the landmass of Asia, turning the Indian Ocean into a vast lake. The southern shore of this lake was designated part of Terra Australis incognita, and belief in this mythical land persisted until well into the seventeenth century. Ptolemy's great contribution

to geography lies in the fact that he preserved and transmitted the works and theories of the ancients and thus established a framework which inspired the later European explorers.

Ptolemy was also a mathematician and astronomer, the last great astronomer of antiquity, and he achieved other important work. He studied mostly in Alexandria and his observations made there, extending the work of Hipparchus, resulted in what came to be known as the Ptolemaic System. This geocentric cosmological theory sited the earth, motionless, at the centre of the universe, with all the heavenly bodies revolving around it; Ptolemy retained the calculations of the Greek philosopher Pythagoras (c582–c507 BC) to explain the movements of the planets against the backcloth of seemingly fixed stars.

Ptolemy's writings on astronomy and geography were the standard textbooks until the sixteenth century, when his theory was superseded by the system of Nicholas Copernicus (1473–1543), the Polish astronomer. This was the first European heliocentric theory of planetary motion, which placed the sun motionless at the centre of the solar system, with the earth and all the planets revolving around it.

The maps of the post-classical European world were simplistic and dominated by Christian biblical tradition and by the teachings of the Church fathers, in which science and observation played little part: the world was divided into three parts, representing the shares of Noah's sons after the Flood. Isidore of Seville (c570–636), a Spaniard of noble birth and great scholarship, who was at one time Bishop of Seville and had sainthood conferred upon him, produced what proved to be a compromise between classical and medieval learning. He posited a divided sphere with Asia at the top and the lower part split into Europe and Africa, themselves divided by the Mediterranean Sea and the rivers Nile and Don; but he still believed that the earth was a globe. Other medieval theologians conceived of the earth as a large box, with heaven as the lid.

Cartographic and religious theories culminated in the so-called Mappa Mundi, a notable example being that of about 1280 preserved in Hereford Cathedral in England. In these maps, the location of Jerusalem and the Holy Land, the focus of the Crusades, and the location of Paradise became crucial.

More practical mariners used charts called portolanos, which were based on practical experience gained on voyages. In these they were greatly assisted by the compass, which had arrived in Europe from the East in the twelfth century; prior to that, navigators had to rely on the Pole Star. Although thousands of portolanos must have been made, because they were for purely practical use not one has survived from before about 1275.

These charts were often gathered into atlas form, and a particularly sumptuous set dating from about 1375, made for King Charles V of France, is known as the Catalan Atlas. In this, west Africa is identified as a source of gold, which acted as a lure for later European explorers.

Although Ptolemy's writings were forgotten in Europe until late in the fifteenth century, they were preserved and even improved upon by such Muslim scientists as al-Idrisi (1070–c1165) who lived in Palermo at the court of the Norman King Roger II of Sicily.

In the following century a Byzantine monk named Planudes (c1260–c1310), a man of great learning, sought and found in Constantinople a text of

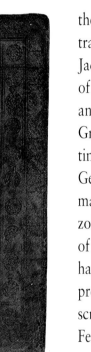

Ptolemy's *Geographia*, for which he commissioned a new set of maps. Planudes' writings were voluminous and his most influential work, *Etymologies* or "Origins", an encyclopedia of all human knowledge at the time, provided scholars of the Renaissance with a good measure of classical learning, although it was a derived work.

A copy of Planudes' text for the *Geographia* was given to the Byzantine emperor, and in 1395 this was brought to Italy as a gift by Manuel Chrysoloras, a diplomat, who was seeking aid against the Turkish threat. Chrysoloras is renowned as the writer and teacher who introduced Greek literature into Italian renaissance cities, notably Florence. He also translated the works of Plato and Homer and, through his teaching, the culture of classical Greece became the foundation of humanist studies in the West.

In 1406, the text of the *Geographia* was translated into Latin by Jacopo d'Angelo, one of Chrysoloras's pupils, and dedicated to Pope Gregory. The Benedictine monk Nicholas Germanus redrew the maps on the trapezoidal projection (one of three which Ptolemy had proposed), and he presented the manuscript to the Duke of Ferrara in 1466.

It was this copy of the *Geographia* that became the basis for the first printed Ptolemy atlas, published at Bologna in 1477 in an edition of 500 copies.

The Ptolemy of 1482, *below*, although crudely coloured, is well printed and great skill is evident in the tooled leather bindings of the 1482 and 1490 editions, *above*.

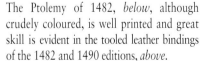

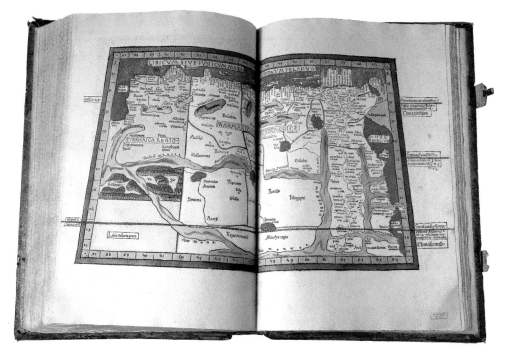

North Africa, from the Gulf of Sirte to the Red Sea, one of four maps of Africa in the atlas

THE GRECO-EGYPTIAN astronomer and geographer, Ptolemy, was active at Alexandria between AD90 and 168. His *Guide to Geography*, better known as the *Cosmographia* or *Geographia*, was compiled in about 160. The surviving manuscripts of the work contain maps, but the earliest of these date only from the late thirteenth century. Nevertheless, since they were compiled from the coordinates for the 8,000-odd places recorded by Ptolemy, they reflect what he would have included in his charts. A copy of Ptolemy's text, in Greek, was brought to Italy in 1406 and translated into Latin by Jacopo d'Angelo.

The translation was published, without maps, in 1475 and, with one exception, the same text was used for another six editions of the *Geographia*, illustrated with maps, that were produced before the end of the fifteenth century. The first illustrated edition appeared in Bologna in 1477; all the maps were drawn on the conical projection—one of the two presentations devised by Ptolemy—by Dominus Nicholas Germanus who was active in Italy in the 1450s and '60s. The Ulm edition of 1482 is the most elaborate and some contemporary maps were added to it for comparison with the earlier ones.

The first page of the 1482 Ulm edition of Ptolemy's *Geographia* is illuminated with a picture of Dominus Nicholas Germanus, a Benedictine monk from the abbey of Reichenbach in Germany, presenting a copy of the work to Pope Paul II.

Nicholas Germanus devised the trapezoidal map with converging meridians and curved parallels that was first used in the 1478 Rome edition and became typical of printed Ptolemaic atlases. He also reduced the size of the maps so that they would fit easily into "the common size of books".

The capital "C" of the word Cosmographia contains a fanciful portrait of Ptolemy, who holds a sphere marked with curved parallels and a pair of dividers.

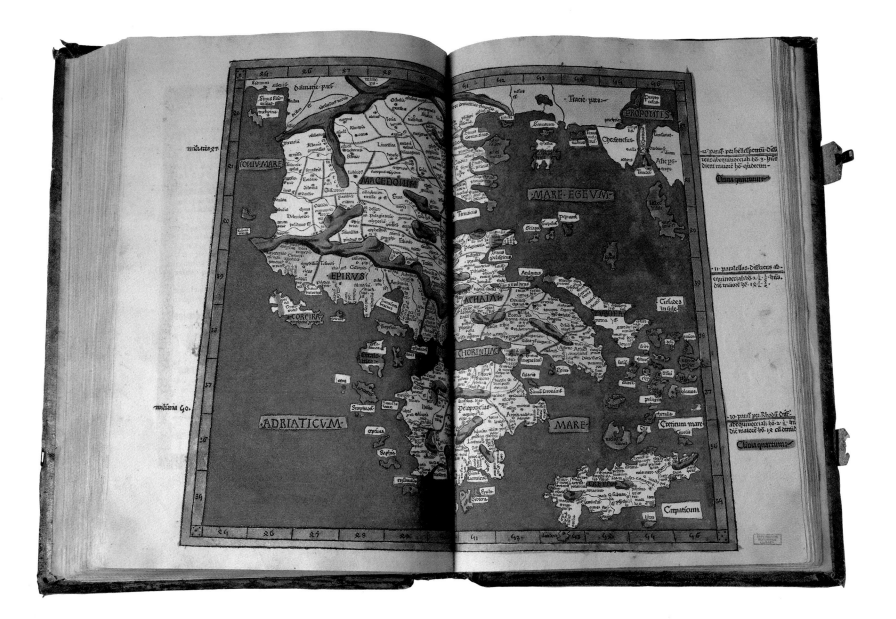

The striking expanse of blue sea in this map of Greece and the Aegean islands is effective, but the use of flat colour was probably due to an inability to introduce shading on the wooden printing blocks to distinguish land from sea. Although copper plates were used to print maps in earlier Ptolemys and were revived in the 1500s, the Germans abandoned them in favour of wood blocks, possibly because of a lack of skilled craftsmen.

This map of the Holy Land has been turned on its side so as to fit the pages of the atlas. Thus the Sea of Galilee, the Lake of Tiberias and the Dead Sea appear left to right, rather than top to bottom.

Places mentioned in the Bible are identified, and information is given about many of those associated with the Old Testament. The curved parallels at the top of the map indicate the projection used.

13

Many woodcut initial letters, intricately embellished and hand coloured, adorn the text of the atlas. The dropped capital "M", *above*, introduces the name of Marinus of Tyre, who is credited with inventing the plane, or cylindrical, projection, in which the lines of latitude and longitude are depicted as a squared grid.

This celebrated and highly decorative map of the world states, in the top border, that the wood block was cut by Johannes of Armsheim, a town near Mainz in Germany. Drawn on Ptolemy's homeotheric projection, which entailed the representation of both latitude and longitude as curved lines, the world is surrounded by fair-haired Teutonic cherubs puffing the winds. The map is so visually attractive that it became extremely popular; even today some 120 copies of an edition of less than 1,000 survive.

The capital "D" is typical of the illuminated capitals in the atlas, most of which are decorated with similar stylized floral motifs.

The 1482 edition of Ptolemy includes Nicholas Germanus's latest thinking, and the world map is extended toward the northwest to include more of Scandinavia; it is the first to show Greenland and Iceland. None of the available information about the west coast of Africa is included, but the Canary Islands, then known as the Fortunate Islands, appear.

The Tropics of Cancer and Capricorn are marked, as is the Equator, although not correctly placed. India's shape was misunderstood: it was depicted as a series of islands. The great rivers Indus and Ganges are shown, but the size of Sri Lanka is much exaggerated, perhaps because of its importance as a trading centre. Mountains are shown as brown areas and many names are those of Roman provinces, such as Dacia, whose modern name, Romania, recalls its historical connections.

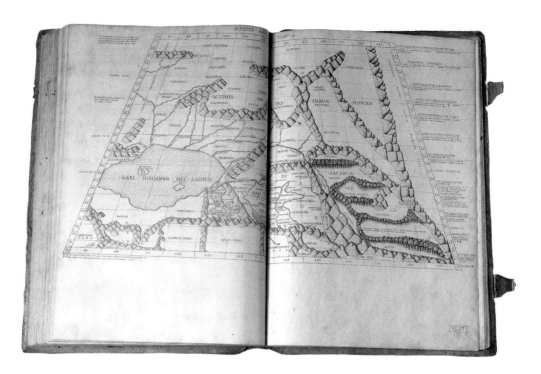

Central Asia, east of the Caspian Sea

Arrian's account of Alexander the Great's exploits, is the source for the map above. Note the detail around Bactria, the rivers Jaxartes and Oxus and the mountain ranges his soldiers crossed.

THE 1490 PTOLEMY, printed in Rome, is radically different in appearance from the Ulm edition, for the Italian Renaissance was exerting a profound influence on printing, and ancient Roman inscriptions, once more esteemed, affected type design. The 27 maps are from the 1478 Rome *Geographia*, planned by the German printer Conrad Sweynham, who developed ways of printing from copper plates. It was famed for its majestic map inscriptions which were punched on the plates, not engraved, but since the maps were engraved, seas and mountains could be indicated by shading.

This map of Europe includes part of northern Italy. Venice, merely marked on the printed map, has been embellished with a hand-drawn view of the city and ships on the lagoon.

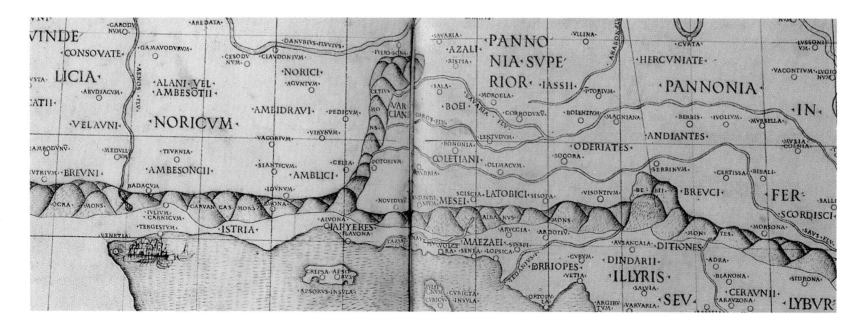

The island of Sri Lanka, or Taprobana—a name derived from the ancients' Taprobane, meaning copper leaf—features prominently in this detail from the world map, *below*. Sri Lanka was a vital trading link between East and West, and both Arab dhows and Chinese junks plied the routes across the Indian Ocean by short hauls between trading posts. Many of the Muslims on the island today are descendants of early Arab traders. The trade routes can be traced by identifying the source of early artefacts, but after the 8th century they are in part documented by Arab navigation manuals and geographical accounts.

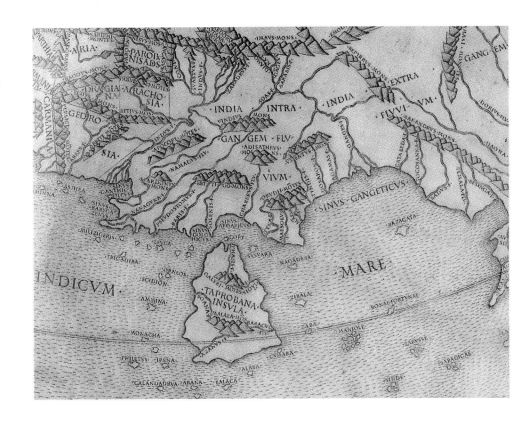

The 1490 world map, although it appears quite different from that in the 1482 Ulm edition, is almost identical in content, since both editions used the same basic material of Nicholas Germanus, the Benedictine monk whose maps were first published in 1478.

PTOLEMY:
GEOGRAPHIAE 1508

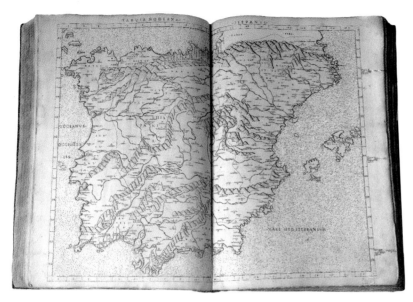

The Iberian Peninsula, with the Roman provinces with which Ptolemy was familiar

T HE PTOLEMY ATLAS of 1508, published in Rome by Bernardus Venetus de Vitalibus, is the first edition of the *Geographia* to include European voyages of exploration to the New World. The atlas is also notable for the inclusion of the world map of Johan Ruysch, a Flemish cartographer who may himself have sailed to America. Much of the material in this world map is either derived from the source used by Giovanni Contarini for his world map of 1506, engraved in Florence, or was based on it.

On Contarini's map, the West Indies are carefully charted and a mere 20° of sea separates them from Zipangu (Japan). North America appears only as a promontory to the north of the islands, but South America, here designated Terra Sancte Crucis, was better understood. Contarini's map was so designed that it could be folded as a cone, which then displayed the "whole spherical world combined into 360°"; Ruysch's map is on a similar projection but includes some different information.

The putto in this "Q" holds what seems to be a carnival mask on a stick, and another mask appears on the shield.

In Ruysch's map of the world, the north coast of South America is well defined and some West Indian islands, such as Hispagniola (Haiti), are marked, although half of Cuba is omitted. The inscription at the unexplored west end of the island reads, "The ships of Ferdinand, King of Spain, reached here." The landmass of eastern Asia is depicted and China is identified by a city marked as Cathaya, but Marco Polo's Zipangu (Japan) is strangely absent. This map, and Contarini's, show what Vasco da Gama had proved: ships could sail around Africa to reach the Far East.

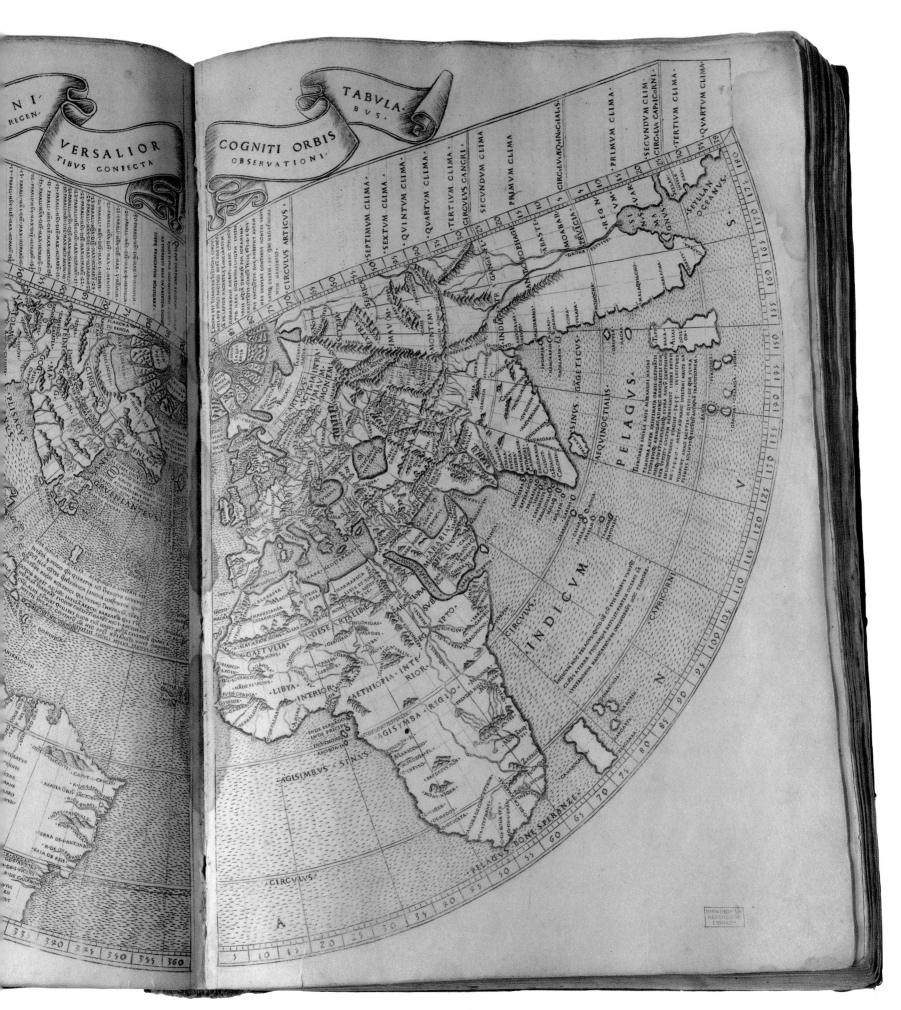

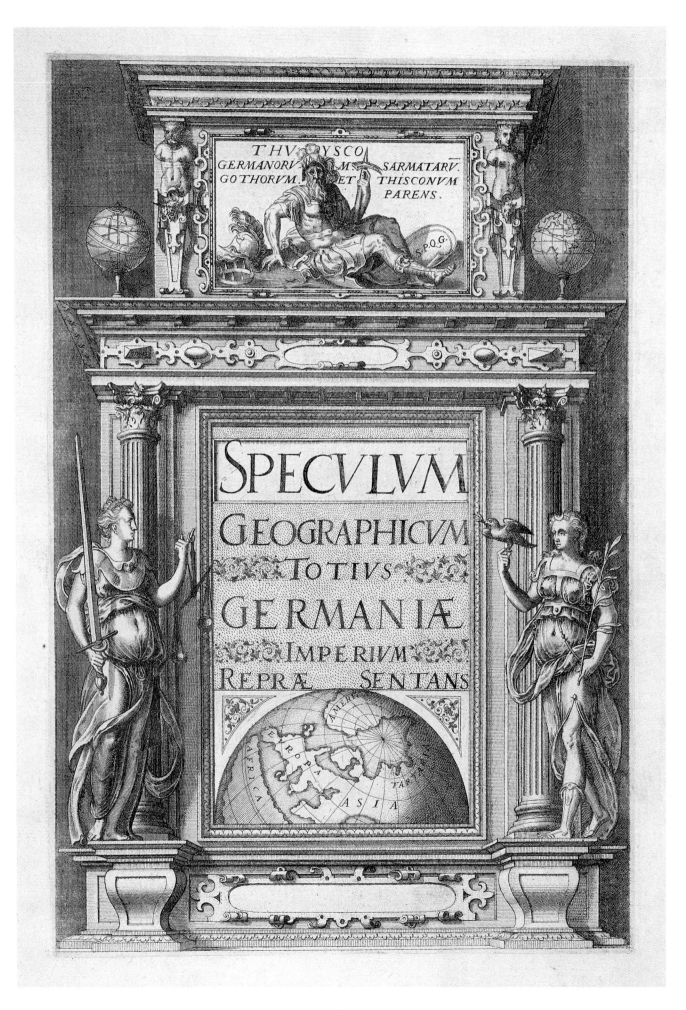

CHAPTER TWO

The ERA of
EXPLORATION

THE RICHNESS, SCOPE and variety of the atlases produced during the fifteenth and sixteenth centuries reflect and record the ever-widening horizons of their European originators. Previous civilizations relied on dogma, inherited tradition, speculation and theory.

With the sciences of astronomy and mathematics in their infancy, the maps of the fifteenth and sixteenth centuries evolved slowly and were largely based on information supplied by sailors, merchants, pilgrims and other travellers. The wonder is not that they are often inaccurate, sometimes wildly so, and that some material is fabricated, but that, given the primitive technological resources available to cartographers and the paucity of reliable information, they invariably provide instantly recognizable images. They lack the accuracy of the modern road map, but have a venerable splendour of their own.

At the beginning of this era of exploration, overland trade to the East had already made the fortunes of Italian city states, such as Venice, which had a dispensation from the Pope to trade with the Muslims. In 1453 the Turks finally conquered Constantinople and closed the overland trade route to the East, so the Europeans lost access to such

Although unsigned, the title page, *above*, to the Ptolemy of 1522, is probably the work of Albrecht Dürer.

The section on Germany in Jode's 1578 atlas is introduced by a finely engraved plate, *opposite*. The globe has been tilted to show Europe and much of the northern hemisphere, and a martial figure personifies the country.

luxuries as precious metals and silks and to the supply of spices so essential for their diet and for the preservation of food.

In order to regain access to this lucrative trade and also to cut out the Eastern middlemen, the Portuguese, under the direction of Prince Henry the Navigator, were determined to find a sea route to the East. From 1419, therefore, Henry ordered his ships to explore the West African Coast, for he believed that from there it would be possible to take a short cut eastward across land to the Indian Ocean and then to follow the established sea routes to India and the Far East. According to the maps of Ptolemy, there was a land link with India, the Sinus Hesperus, from the lower part of Africa to a point somewhere beyond the subcontinent, creating a landlocked Indian Ocean.

The Portuguese already held key positions in North Africa and had been deterred from moving south overland only by the barrier of the Sahara Desert. Henry hoped to find Prester John, a mythical Christian ruler, to fight with him against the Muslims. He was disappointed in this, but his men did eventually find gold at Mina on the Gold Coast.

After Henry's death in 1456, the Portuguese continued to explore the West African Coast.

Between 1469 and 1474, the ships of the wealthy merchant Fernaõ Gomes from Lisbon opened up another 2,000mls/ 3,200km of the coastline. When João II became king of Portugal in 1474, he took control of the exploitation of Guinea.

In 1482 Diogo Cão sailed 850mls/ 1,370km across open sea to Cape Lopo Goncalves, relying on improved navigational instruments—the astrolabe for measuring latitude by the height of the sun or stars, a superior compass, and sandglasses for measuring time over distance, since no watch or clock had yet been invented that was accurate over a long period of time. He probably also had access to the manual of navigation that had already been compiled by the king's scientific advisers but was not published until 1495. Diogo Cão added another 1,000mls/ 1,600km of coast to the possessions of the Portuguese Crown.

In 1488, Bartolomeu Dias at last succeeded in rounding the Cape of Good Hope, more suitably named the Cape of Storms by Dias himself. His success was ultimately due to his wisdom in taking an extra store ship so that his crew could survive the desert regions of southwest Africa.

Disproving the existence of Ptolemy's land link, he sailed up the southeast coast of present-day South Africa and reached the Great Fish River. The map of Africa in printed atlases was not shown complete until the Ptolemy of 1508, where the Cape of Good Hope is marked as C. de bona Spe.

There was a setback for Portuguese expansion in 1492, however, when Christopher Columbus, the Genoese navigator, who had long lived in Portugal, changed his allegiance to King Ferdinand and Queen Isabella of Spain and, with their backing, sailed west across the Atlantic

Fine leather bindings, ornately tooled and blocked in gold and bearing clasps, *above*, were common on early atlases, reflecting their importance and value .

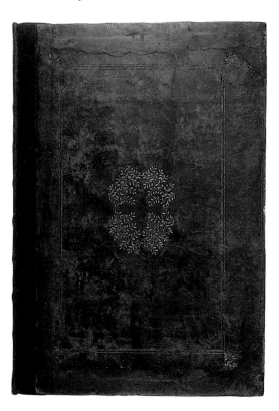

Ocean in an attempt to claim the profitable spice trade for Spain. By chance, he discovered the West Indies and the New World instead.

He had miscalculated the distances involved, misled because he trusted optimistic calculations of the Arabian mathematician Alfragenus, rather than those of the ancient Greek Erastosthenes, and he gravely underestimated the width of the Atlantic. Columbus never realized that he had discovered a new continent and always believed that the West Indies were islands off the coast of Asia, although toward the end of his life he did concede the existence of a great southern continent.

The first manuscript map of America was made by Columbus's pilot, Juan de la Cosa, in 1500. The first printed map was that of Contarini in 1506; a further printed map showing America, of which 1,000 copies were sold, was published by Martin Waldseemüller in 1507. The world map in the Ptolemy of 1508 shows America as the small islands of the West Indies, with no realization of the great continent beyond.

Waldseemüller's map of the new discoveries, in his Ptolemy of 1513, shows greater detail and promotes Amerigo Vespucci's false claim that he was the first to discover mainland America by naming the continent after him. This atlas had, also, two up-to-date maps of Africa, with the names of all known coastal places.

In 1493–94, the whole of the newly discovered world was apportioned between Portugal and Spain by a bull, issued by the Spanish Pope Alexander VI, known as the Treaty of Tordesillas. An imaginary line was drawn 370 leagues (about 1,000 mls/ 1,600 km) west of the Cape Verde Islands off the coast of Africa. Spain was given sovereignty over all lands to the west of it,

Portugal over those to the east. Thus Spain received all of the New World, while Portugal retained its control over Africa and the adjacent islands.

Portugal had not, however, given up hope of winning the sea route to the East, and in 1495 Vasco da Gama (c1460–1524) made preparations for a voyage to reach India. He followed Dias's route around the Cape of Good Hope and continued up the east coast of Africa, via Mozambique, to Mombasa in East Africa, where he engaged a local pilot who knew the sea routes across the Indian Ocean. In 1498, he finally made landfall at Calicut on the southwest coast of India. Da Gama's connection with the subcontinent continued throughout his life, first as adviser on India to the Portuguese kings and finally as Portuguese viceroy at Goa.

From England, John Cabot, under the patronage of Henry VII, a newcomer to empire building, sailed west from Bristol in 1497 and reached the rich fishing grounds of Newfoundland. Like Columbus, he believed he had reached the East Indies. In a second voyage, he hoped to reach Japan but instead found Labrador.

Portuguese empire building continued to flourish when, in 1520, Pedro Cabal sailed west, still trying to reach India, and discovered Brazil. He claimed the whole of this southern continent of America for Portugal, even though it lay west of the dividing line established by the Treaty of Tordesillas.

Fernaõ de Magalhais (known as Magellan), also Portuguese, altered the balance of power again when, like Columbus, he changed his allegiance to Spain. Convinced that he could reach the Spice Islands by sailing westward, Magellan succeeded in persuading the emperor Charles V to fit out an expedition. In 1519, with five ships and a crew of mixed nationality, he sailed across the Atlantic, down the east coast of South America, through the stormy Magellan Straits and into the Pacific. He reached the Philippines in 1521, after a gruelling voyage marked by hardship and mutinies. Magellan was killed in a skirmish with the natives on

Mactan Island, and just one ship, under Elcano, returned home via the Cape of Good Hope, carefully keeping south to avoid the Portuguese. Magellan's companions had proved that the earth was much bigger than any of their contemporaries had believed, although in ancient times Erastosthenes had deduced the distance correctly.

Meanwhile, also in 1519, in the Americas, Hernan Cortes from Spain conquered the Aztecs of Mexico and destroyed their great cities. Those people who were not killed died from diseases brought by the invaders. Cortes had been attracted by the Aztecs' gold and he and his soldiers—the Conquistadors—divided and plundered the Aztecs' land and destroyed the strong administration and established agricultural systems. Now there was a vast market for the African slaves, shipped across the ocean in *tumbeiros*, or coffin ships, by the Portuguese to work the land in America.

The French joined in the voyages of discovery when their king, François I, despatched Giovanni da Verazano to look for a westward passage to the Spice Islands. In 1522–28 he sailed the American coast from Florida to Newfoundland and was, possibly, the first European to enter New York Bay: the bridge across the narrows at the entrance to New York harbour is today named after him. By his exploits he proved the size of the North American continent, but he failed to find a sea lane through it. Jacques Cartier's voyages in 1534–42, exploring the American east coast and the

The delightful picture of camel-drawn carts, sheep and frisking horses, *above*, appears just north of the Caspian Sea on a map of Russia in Jode's atlas of 1578.

St Lawrence River, laid the basis for later French claims to Canada.

By 1580 much significant information, brought back from all over the world by explorers and navigators, had been incorporated into the atlases of the period as, for instance, the volumes published by Antonio Lafreri demonstrate. In less than 100 years, European explorers had discovered North and South America and greatly extended their knowledge of Africa, India and the Far East.

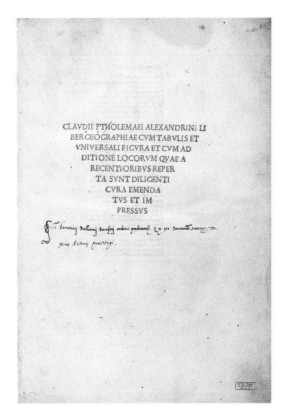

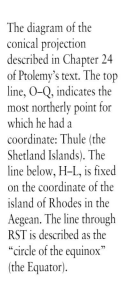

THE *GEOGRAPHIAE* OF 1511, printed by Jacobus Pentius de Leucho, was published in Venice and contained 28 maps. Ptolemy's text was edited by Bernardus Sylvanus of Eboli, who was the first to question and amend the great cartographer's words, rather than slavishly reprinting them. This was a significant step, for many of Ptolemy's opinions were erroneous and much practical information gathered by navigators was frequently discounted because it could not be reconciled with Ptolemy's ideas.

The title page of the atlas records that many additional places which were "recent discoveries" are marked on the maps and that the text had been "diligently amended and edited". The use of red ink on the title page is also an innovation, and this is the first edition to combine red and black in the printing of the text and of the actual maps. It is interesting to note that this copy of the atlas originally belonged to a member of the order of preaching friars, who inscribed the title page.

Title page to the *Geographiae* of 1511

The diagram of the conical projection described in Chapter 24 of Ptolemy's text. The top line, O–Q, indicates the most northerly point for which he had a coordinate: Thule (the Shetland Islands). The line below, H–L, is fixed on the coordinate of the island of Rhodes in the Aegean. The line through RST is described as the "circle of the equinox" (the Equator).

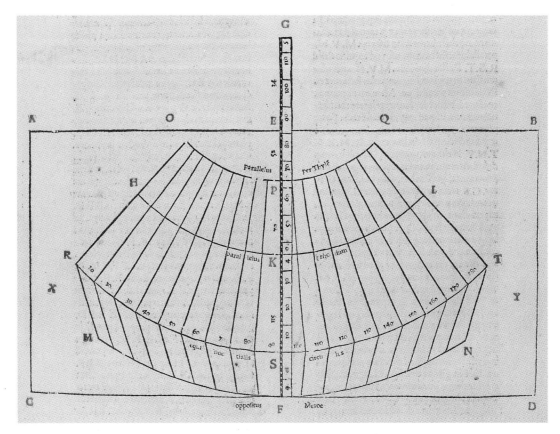

The two borders, *above*, are from the "Eleventh Map of Asia" and the "First Map of Africa", which is decorated with two parrot-like birds.

France and Flanders are bounded on the northeast by the River Rhine and in the southeast by the mountains of Switzerland, in the Third Map of Europe, *below*. The rivers Rhône, Seine and Loire are shown, but no other identification is given save, in red, the names of the tribes that occupied the region in Roman times and the main Roman settlements.

The border to the map of Asia, which embraces northern India, appropriately displays two tigers, creatures that are found only there.

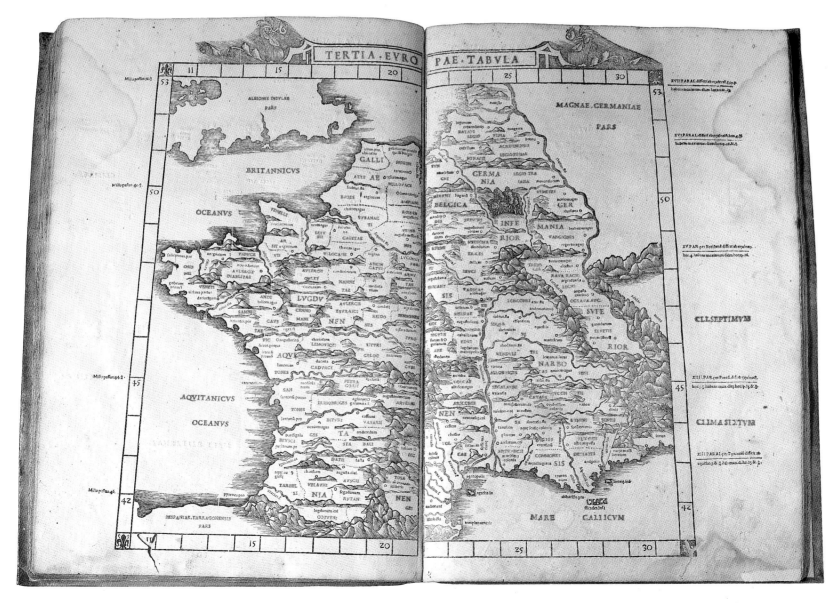

The great importance of this map of the world in the history of cartography lies in the cordiform projection, used here for the first time. The map continues the practice, established in the 1508 edition, of including information about areas of the world which were currently being explored.

Unlike the 1508 version, however, this map shows the whole of Cuba and much of Brazil, under the title Terra Sanctae Crucis. North of Cuba and Haiti, parts of North America, labelled Regalis domus and Terra Laboratorum (Labrador), appear. This is the first map to include graphic representations of the discoveries of the Portuguese Corte-Real brothers in 1500.

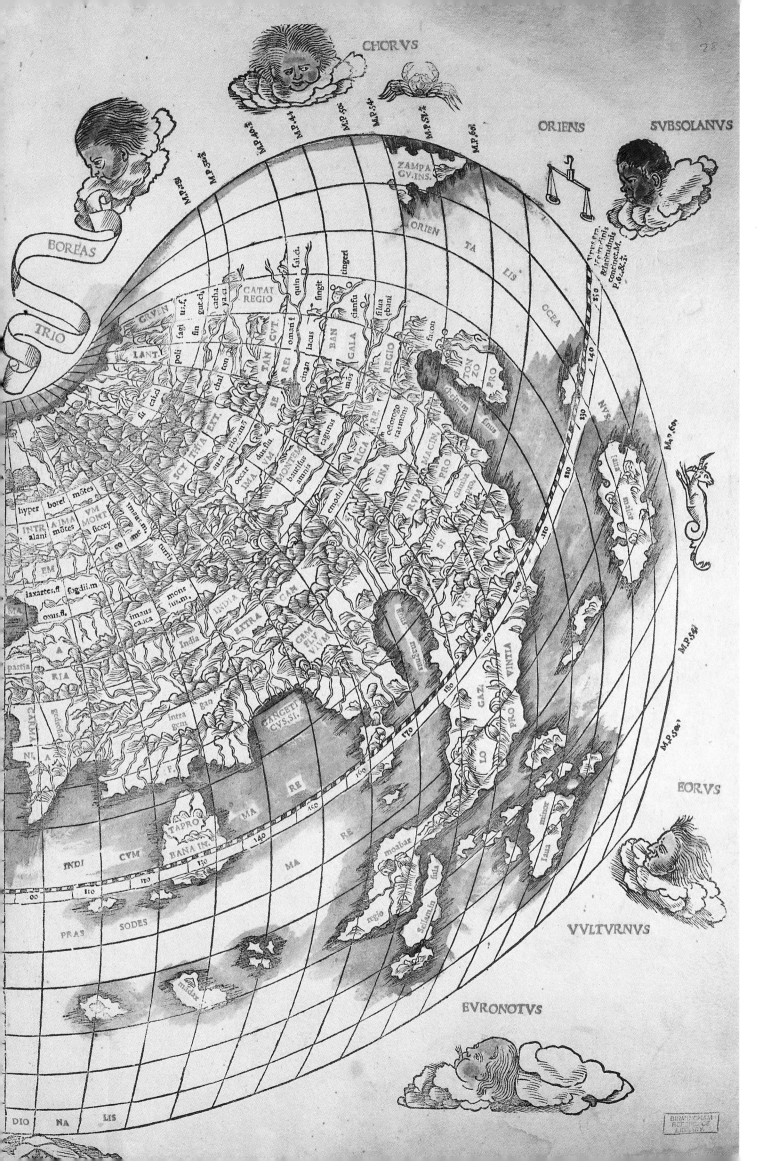

On the other side of the world, in the Orient, the map depicts the island of "Zampagu", Marco Polo's Japan. It is impossible to identify with certainty the two islands named Java Major and Java Minor, but the island incorrectly placed to the east of the Malabar coast is probably Sumatra, and the island to the south, Java.

In early 17th-century maps, Java is named, but Java Major, which was mentioned by Marco Polo, may refer to New Guinea or even allude to the existence of Australia.

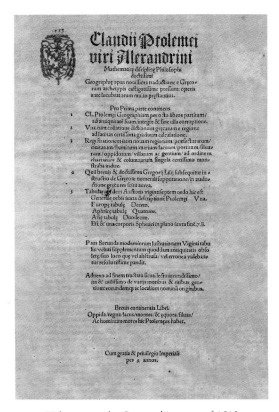

AFTER 1490, no new edition of Ptolemy's *Geographia* was produced until 1507, but that was reprinted a year later with the addition of American material. More American material was added to the important Venetian edition of 1511, but the best was the later edition from Strassburg in 1513. The 20 new maps in this printing were compiled by Martin Waldseemüller, a canon at the church of St Dié in the Vosges mountains and a notable cartographer.

Title page to the *Geographiae opus* of 1513

Waldseemüller culled material not only from the libraries of the nearby universities of Basel and Strassburg, but from up-to-date reports provided by Spanish and Portuguese navigators. The new maps included the first devoted solely to the coast and islands of America—the New World. Waldseemüller's atlas, for such the *Geographia* had virtually become, was the finest and most accurate collection of maps assembled until that published by Ortelius some 60 years later.

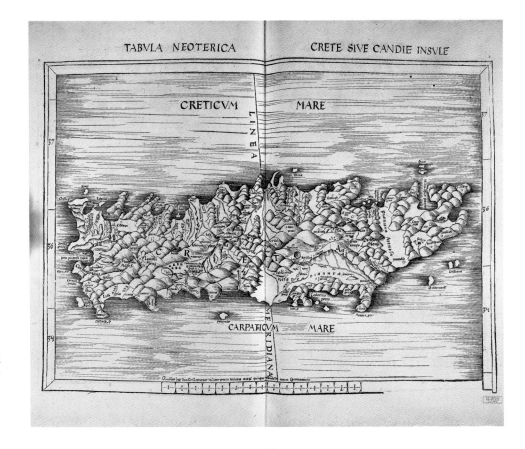

The depiction of Crete is typical of the detailed local and regional maps that publishers were adding to the original Ptolemaic collection.

The amount of information given indicates the skill and diligence of both surveyor and cartographer, for the accuracy of the coastline and the general topographical detail are remarkable.

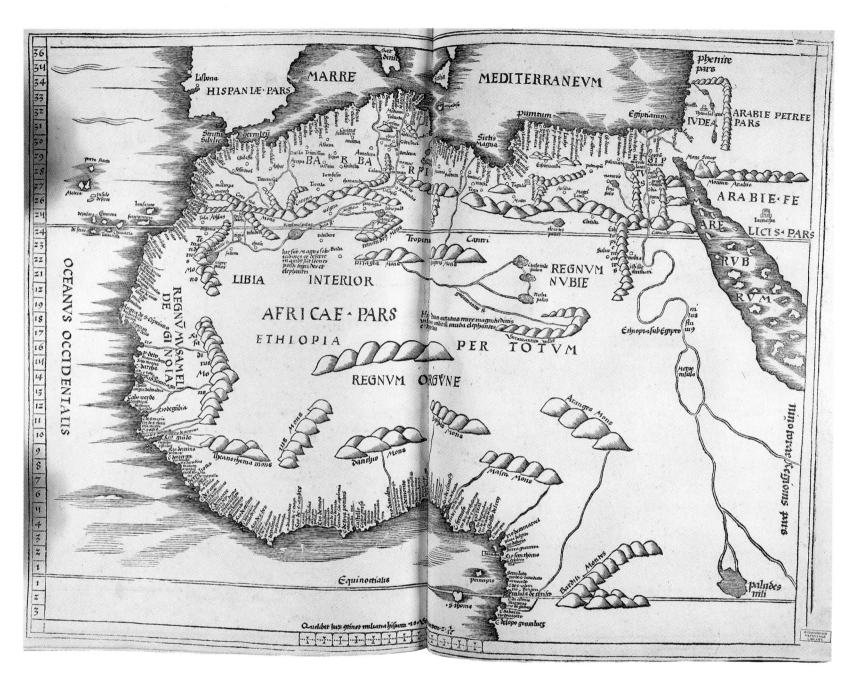

The two "modern" maps of Africa provide a radically different image of the continent from that in earlier Ptolemys. By the 16th century, navigators had explored and charted the coastline with considerable precision, and every significant bay, inlet, river and settlement had been noted. Most importantly, the shape of the landmass is, for the first time, truly understood.

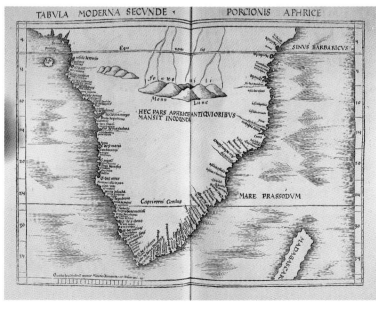

The mispositioning of the island of Madagascar mars this otherwise accurate map of southern Africa. This often occurred with the mapping of islands, for navigators charting the continents could stay close to the shore, but when they sailed out into the open ocean, their accuracy in charting declined markedly.

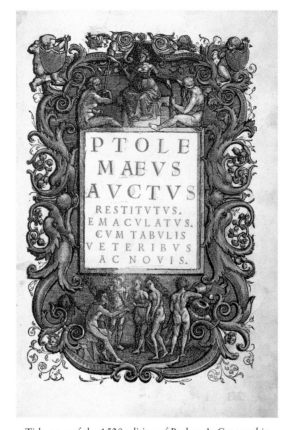

THE 1520 EDITION of Ptolemy was a reprint of the Strassburg edition of 1513, as far as the maps were concerned, but the text was reset, and this elaborate, hand-coloured woodcut title page was added. The title claims the book contains "both old and new maps", but fails to point out that nothing has actually been added.

The advent of Ortelius's *Theatrum orbis terrarum* rang the death knell of the Ptolemy industry, although atlases continued to be produced in a desultory fashion for another 160 years, with further editions of the *Geographia* containing also local maps and new maps, which were added as a result of the explorations by European navigators.

Title page of the 1520 edition of Ptolemy's *Geographia*

The woodcut map of the region around Strassburg, where this atlas was published, was printed from the same block as that used for the 1513 edition. It was probably included in order to ensure local sales.

West is at the top of the plate, thus Frankfurt appears at the bottom right, Basel near the top left, and Argentina, the Latinized name for Strassburg, in the centre. The mountains of the Vosges and the Black Forest are shown, and the great artery of the river Rhine unites the area.

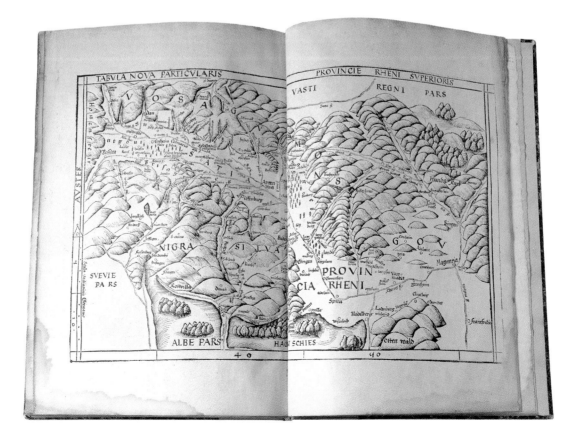

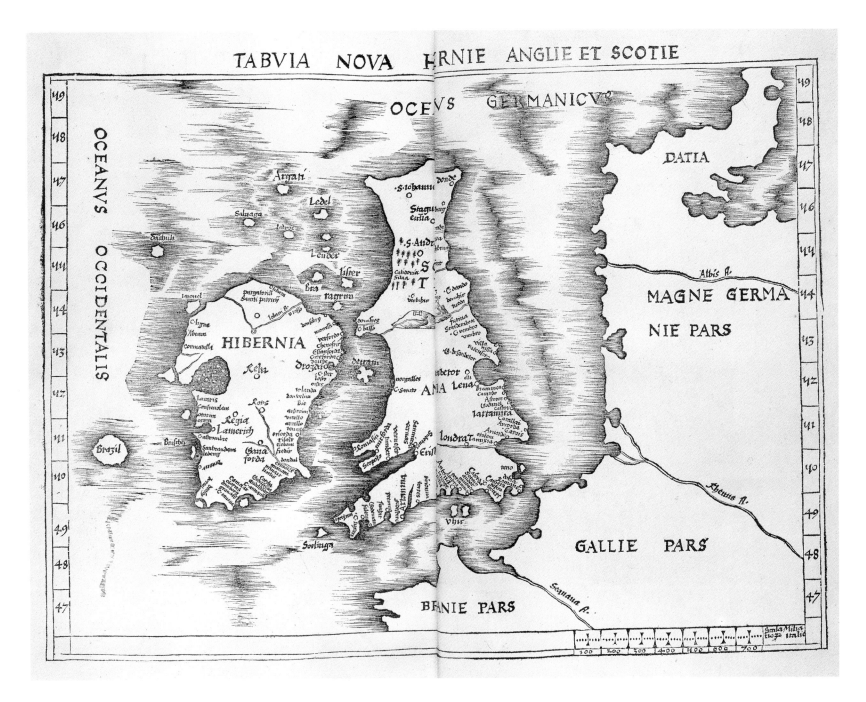

In early editions of the *Geographia*, maps of Great Britain show Scotland twisted eastward, at right angles to England. The 1513 edition corrected this, giving Scotland its true north–south orientation, and this plate shows Britain in recognizable form, although there are many inaccuracies. An interesting feature is the inclusion of a mythical island of Brazil, west of Ireland. The presence of islands in the western ocean was accepted by early navigators, and Brazil appears in the great Catalan Atlas of 1375.

The western isles of Scotland are oddly dispersed, but Ragran is probably Arran, and farther south in the Irish Sea the Isle of Man, in its heraldic shape, is marked.

Northwest of Ireland is Thule (Shetland), the most northerly point for which Ptolemy had a coordinate: here it is placed too far south and west.

It is evident that this map was compiled from mariners' charts, for towns that were important ports are indicated in bold lettering, among them King's Lynn on the Wash (Lena), Bristol (Eristo), Teignmouth (Artanina) and London (Londra).

The only river marked in England is the Thames, but the Elbe, the Rhine and the Seine are shown in Europe, and Ireland (Hibernia) is heavily annotated. In Scotland, the Caledonian Forest, one of the most ancient in Europe, appears—today only a remnant exists.

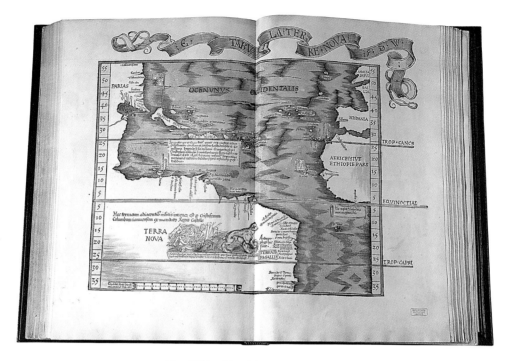

The Old World and the New—Terra nova

The "Alphabetical Register", or index, with its pleasing scrolled heading, listed all the "regions, cities, towns, rivers, mountains" contained in the maps and also the "history attached to them where appropriate".

T HIS ATLAS BY LAURENT FRIES, published at Strassburg in 1522 by Johan Grieninger, contains 50 maps, as against 47 in the 1520 edition. The three additional items are two maps of India, Tabula Moderna Indiae Orientalis and Tabula Superioris Indiae et Tartariae Maioris, and the Tabula Moderna Gronlandie et Russie, which is actually a map of the world, showing the New World. Three other maps also include America, one of which again shows Greenland—as part of Europe. In the map above, part of the east coast of America, south to Brazil, is shown. A label in the middle of a curiously narrow North Atlantic records Columbus's discovery and annexation for Spain of the islands of "Spanoha" (present-day Hispaniola) and "Isabella" (Cuba), marked by a flag bearing the arms of Aragon and Castile. Ptolemy's influence on cartography had begun to wane by the time this edition was published, but atlases using his projection and basic information continued to appear until the early eighteenth century, the last one in 1730.

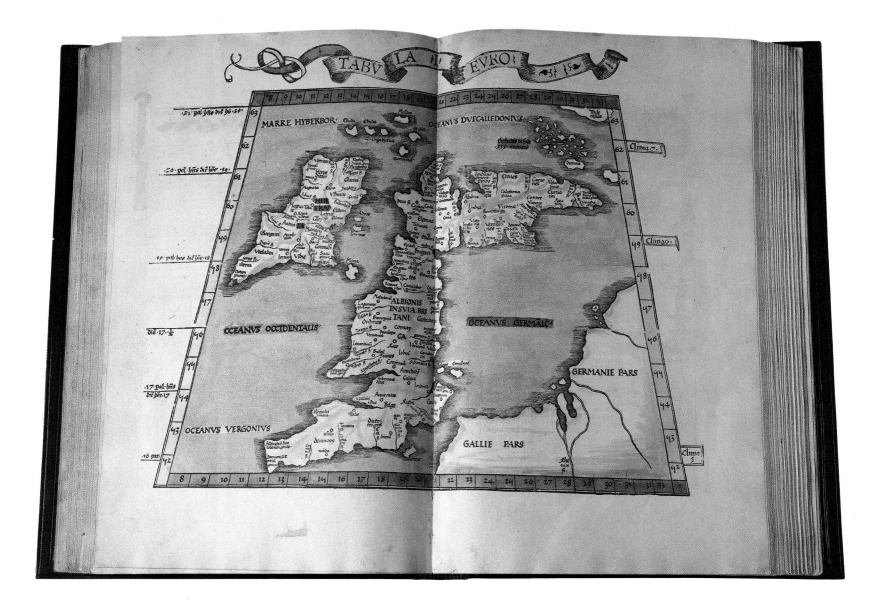

The most striking feature of this map of the British Isles is the depiction of Scotland turned at an angle, although it is easy to recognize the Firths of Forth and Clyde. This oddity arose because Ptolemy rejected the calculations of the Greek astronomer and mathematician Hipparchos, who around 190 BC established the measurement of latitude and longitude by use of an astrolabe, in favour of those of Posidonius (*c*131–51 BC), which were far less accurate.

Ptolemy also decided to continue the measurements of his older contemporary Marinus of Tyre, who tried to produce a map of the world by calculating the latitude and longitude of all known places. Because of these inbuilt errors, Ptolemy estimated the Mediterranean to be some 20° longer than it is, and the longitudinal reading for Scotland thus forced the map maker to show it turned on its side.

The west side of the country is fairly accurately drawn; Ireland has been made to appear smaller than it is, but the Isle of Man, under the name Mona, is marked, as are the Orkneys and the Hebrides. The map is inscribed with the names of the ancient British tribes, and London and York are among the few modern cities that can be identified.

A pair of compasses and a quadrant with a plumb line adorn these capitals.

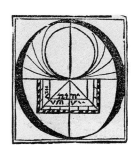

Early printers followed the manuscript tradition of decorated capital letters to introduce important sections in books. In this atlas mainly nautical symbols appear, but the letter "O", *above*, includes the name Jehova in Hebrew characters.

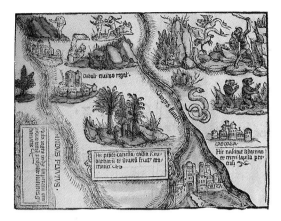

This woodcut appears in the Asian section of the atlas and includes many visual references to the land of the mythical Prester John, supposed in medieval legend to rule over great areas of Africa and central Asia; here his realm is shown between the Indus and Ganges rivers. The deserts north and east of India were thought to be inhabited by bizarre humanoid creatures such as those depicted alongside Adam and Eve in the Garden of Eden.

Lodovico di Varthema, the Italian adventurer, in his *Itinerario* (1510) described the elephants of Taprobane (Sri Lanka) and mentioned their exceptional size. This early depiction of an elephant gives it hoofs like a horse, an ear seemingly growing from its forehead and tusks that point upward. A more accurate representation appeared later, in the Sebastian Münster edition of Ptolemy published in 1540.

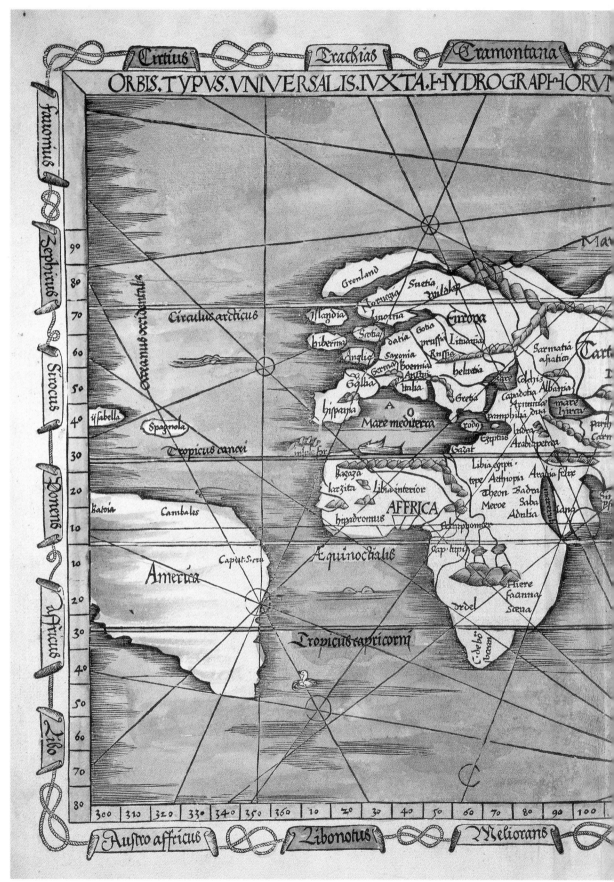

Laurent Fries's rather crudely drawn map of the world is framed by linked panels containing the names of the winds. South America is shown, albeit too close to Africa, and it is given a western coastline. This was inserted more by intuition than by report, for Magellan's ship did not return to its home port in Spain until after the atlas was published. Spanoha and Isabella islands are again correctly marked.

34

Manuscript charts of the seas were made by Portuguese and Spanish navigators from an early date, and they became the prototypes for the earliest printed nautical maps. Portuguese sources provided most of the information for the first hydrographic map to appear in a printed Ptolemy, that of Martin Waldseemüller in the 1513 Strasburg edition.

The slightly later version, *left*, by Laurent Fries, is signed by him and dated 1522. Most maps in the Fries edition are smaller versions of Waldseemüller's but, far from adding information, Fries managed to distort and often to corrupt the earlier work.

The intricate title page to the 1522 edition of Ptolemy, in the North European decorative style, may be the work of Albrecht Dürer, who contributed diagrams elsewhere in the atlas. The symbolism is wholly Christian and Catholic, for at this date cartography and geography had yet to develop an individual symbolic repertory. Saints and cherubs occupy the niches, and the centrepiece depicts the ceremony of the Eucharist.

England and Scotland are separated by a stretch of water (an early error previously eliminated) and Madagascar is sited in the middle of the Indian Ocean. The Malay Peninsula almost touches Ceylon, but the separate existence of Java is carefully recorded. China, under the name Cathaia, is marked but Japan, although mentioned by Marco Polo in his account of his travels, *Il Milione*, written in 1296–98, does not appear.

ABRAHAM ORTELIUS 1570

THE FIRST EDITION of the atlas of Abraham Ortelius (1527–98) was published in Antwerp in 1570 by the famous Plantin press. Ortelius probably began work on the atlas some 10 years before its publication and had published a world map in 1564 and one of Asia in 1567. He has been called the "father of modern geography" and his maps of 1570 certainly constitute the first uniformly conceived atlas. The *Theatrum orbis terrarum* remained in print until 1612 and was translated into the main European languages.

Title page of Ortelius's 1570 atlas

As first issued, the *Theatrum* contained 70 maps—56 of Europe, 10 of Asia and Africa, and one of each of the four continents. The work was dedicated to Philip II, King of Spain, ruler at the time of the Low Countries. The title page, *left*, comprises the usual panoply of classical figures, together with three spheres to indicate its geographical content. Ortelius was a scholar and collector, not a practical surveyor, but he travelled extensively and his friends included the English historian William Camden and Richard Hakluyt, the geographer.

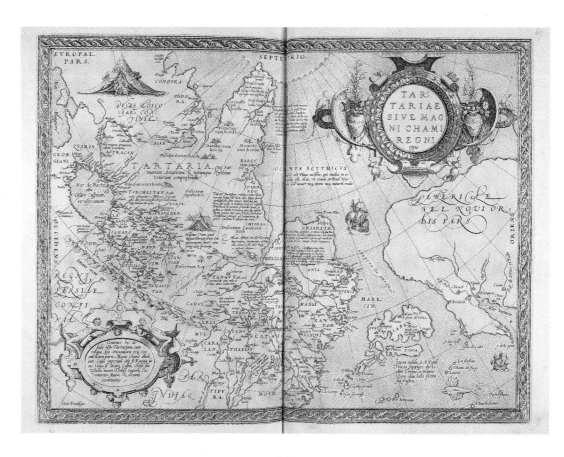

This map of "Tartary or the Kingdom of the Great Cham" alludes to the empire forged by the Mongol, Genghis Khan, in the 13th century. The word "cham" is derived from the Turkish word *khan*: lord or ruler.

Because several Europeans traversed the region east of the Caspian Sea during the Middle Ages, it is well documented, and many places in China are not only identified, but correctly located.

The size of the Pacific Ocean has been wildly underestimated; and although the Straits of Anian, believed to separate Asia from the New World, might represent the Bering Strait, they are too far south.

Ortelius's knowledge of Japan was incomplete. He knew that Honshu was the main island and that Shikoku (Tonsa) existed, but he ignored the northern island of Hokkaido. Osaka can be readily identified.

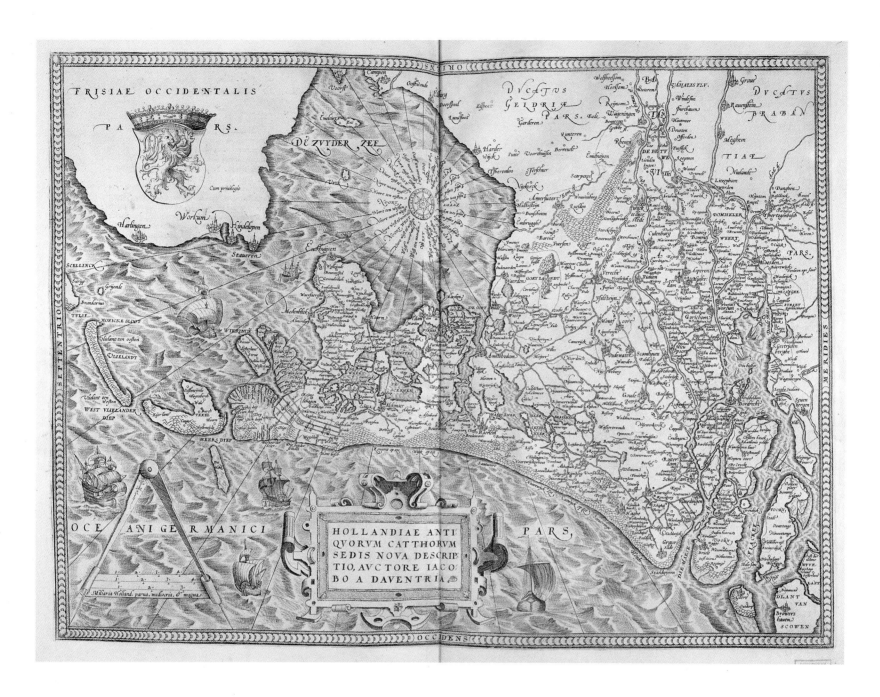

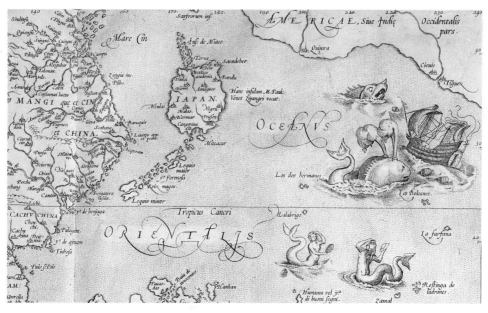

Jacopus Daventria's detailed map of Holland, *above*, is drawn with east at the top. On the seaward side the elaborate embanking already undertaken is clearly shown, although lands to the north of Amsterdam and east of Alkmaar were still to be drained.

Texel and Vlieland at the entrance to the Waddenzee are marked, and the island north of Shagen is yet to be absorbed into the mainland.

In contrast to the map of Japan opposite, this map shows Honshu, Shikoku and Kyushu as a single island, but Kagoshima and the city of Nagoya (Negru) are identified in more or less their correct positions.

The string of islands south of Kagoshima are the Ryuku Islands, here named Lequio. At the top of the main island is a city named Miaco, which also appears opposite but cannot be accurately identified.

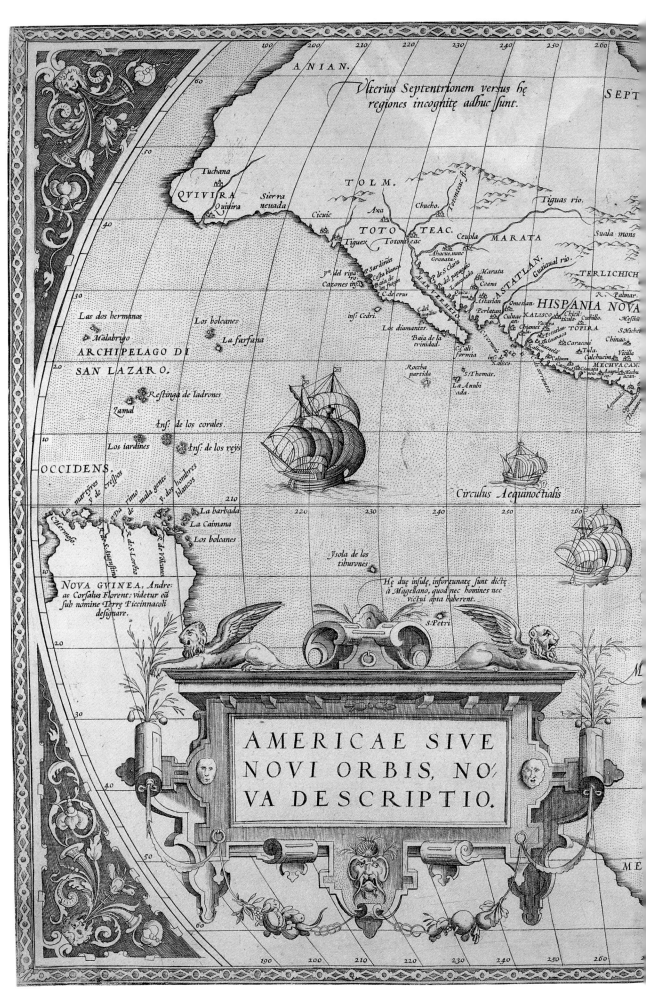

The maps of Sardinia, *above*, and Corfu, *opposite*, appear on one plate in the European section of Ortelius's atlas, as do other Mediterranean islands. The largest of these after Sicily is Sardinia, which in the past had considerable strategic and economic importance. The towns of Cagliari and Sassari are marked, as are many other settlements.

Ortelius's world map included much of the information given in this map of the New World. Here, however, clear water to the north of America is not shown; a curious omission, for Richard Hakluyt specifically asked Ortelius to include material on North America for English navigators, who were at the time searching for a northwest or northeast passage to the East. The shape of North America is distorted because of the projection used and because the interior of the continent was uncharted, but California now appears as a peninsula and part of the landmass, rather than as an island.

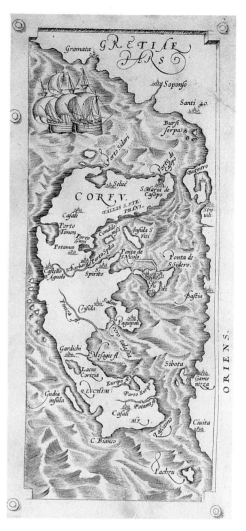

The outline of Corfu, the largest of the Ionian islands, is reasonably accurate, but the citadel of Kerkyra appears larger than it was. Both the harbour (*mandaki*) and the strip of land linking the old fort to the main island are marked.

The battlelines for colonial expansion are already established in North America: Hispania Nova straddles Mexico and Texas, and Nova Francia occupies part of the United States and Canada. North of this is Terra Corterealis, which commemorates the explorations of the brothers Gaspar and Miguel Corte-Real in 1500–01. It is interesting that Canada, the Indian name for the territory north of the St Lawrence River, also appears on the map.

In the Atlantic, the Azores, the Cape Verde Islands and "La Bermuda" are all marked. The shape of the lower part of South America is poorly understood, although the importance of the two major river systems of the Amazon and the Plate is recognized. The Straits of Magellan are marked, but Tierra del Fuego is still linked to Ptolemy's mythical southern continent, and Australia has yet to be recorded.

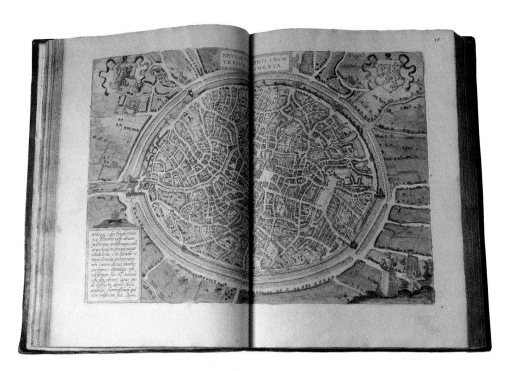

Plan of the walled city of Bruges in Belgium

BETWEEN 1572 AND 1618, a six-volume atlas of hand-coloured plans and views of towns and cities all over the world was published by the Germans Georg Braun and Frans Hogenberg. They probably operated within the circle of Ortelius; Hogenberg certainly engraved most of the 70 maps in Ortelius's own atlas of 1570. The views reflect the early artistic tradition of showing landscapes from a high vantage point—a style that is exemplified by the map of Bruges, *above*.

As was customary in books of the 16th and 17th centuries, the title page of the third volume of the atlas was elaborately decorative. The allegorical style, full of classical allusions, is typical of work from Germany and the Low Countries at the time.

Dependent on the central figure of Justice, with her scales, and the reclining figures of Peace, holding a caduceus—the symbol of a herald—and Concord are the benefits of good order: Plenty, Security, Community and Obedience. At the bottom, by way of contrast, is the Tower of Babel, symbolizing discord.

The Latin inscription beside the map calls Bruges "the prettiest city of German Flanders". But, isolated by the silting up of the Zwijn estuary and overtaken by Antwerp as Flanders' major trading port, Bruges was already past its prime when this plan was drawn. In the 400 years that have since elapsed, remarkably little has changed: the great gate to the left still forms the city entrance, and the central square and town hall remain intact, as does the Stadshuis to the right.

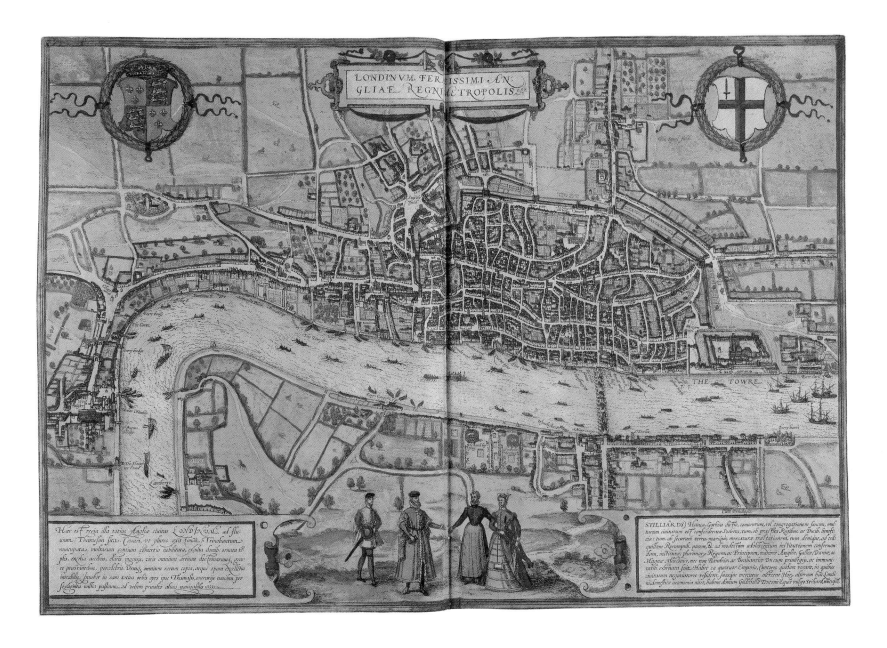

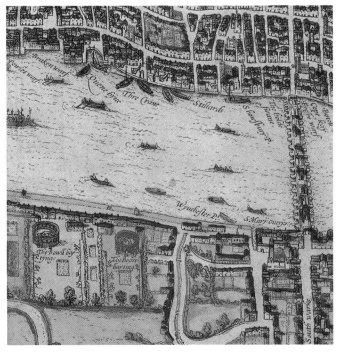

A detail shows London Bridge with houses, shops and a chapel built on it. Until 1750, when Westminster Bridge was opened, it was the only bridge in London across the Thames; elsewhere, river crossings were made by ferryboat, clusters of which may be seen at strategic points. The south bank was only sparsely inhabited, and although there were bull and bear baiting rings the famous Globe theatre in which Shakespeare's plays were performed had yet to be built.

This view of Tudor London shows the city from the moated Tower in the east to Westminster. The city covered scarcely a square mile (2.6² km) and owed its importance to the fact that it was the lowest point at which a bridge could be built across the river; it thus became the controlling point for traffic on the Thames and the country's greatest port.

London was expanding rapidly: between 1530 and 1600 the number of Londoners trebled to around 75,000, with another 150,000 living outside the walls. New industries producing glass and pottery were opening up, often outside the city gates, but London was still small and unimportant compared with other major European cities.

The city of Westminster housed the royal residence. When Henry VIII moved the court to Whitehall Palace after the fall of its builder, Cardinal Wolsey, in 1530, it became the home of officialdom and finally of Parliament itself.

The title pages to
Volumes One and Two
are brilliantly coloured
confections combining
classical architecture
with figures and scenes
that have a religious and
allegorical significance.

In the first, surveyors'
tools are grasped by a
seated woman, perhaps
Athena, the Roman
goddess of Wisdom. In
the second, Cybele, the
goddess of towns, is
drawn by two lions.

When this map, *left*, was published, Paris was one of the world's great cities, with a population of 180,000. It consisted of three main elements: the Ile de la Cité, the site of a pre-Roman settlement on the north–south trade route across the River Seine; the right bank which served as a port and commercial centre; and the left bank where the university lay.

The grid of streets inside the inner wall, built *c*1180–1210, marks the extent of the early city; the great walls beyond were raised in 1337–1442. The cathedral of Notre-Dame is at the eastern end of the Ile de la Cité, to the west is the church of Sainte Chapelle, and in the centre of the plate, the Bastille.

The "View to show the ancient monuments of Rome", *right*, was made in 1570 by P. Ligorio; many of the classical ruins shown still survive. The area in this detail is partly enclosed by the Aurelian walls (*c*280) and by later fortifications to protect the Vatican; the Piazza and Basilica of St Peter can be seen at the bottom left.

The Colosseum is prominent, as are the Pantheon and, above it to the right, Trajan's Column and the statue of Marcus Aurelius on horseback. Piercing the walls at the top is part of the system of aqueducts that brought water to Rome; nearby are the Baths of Diocletian.

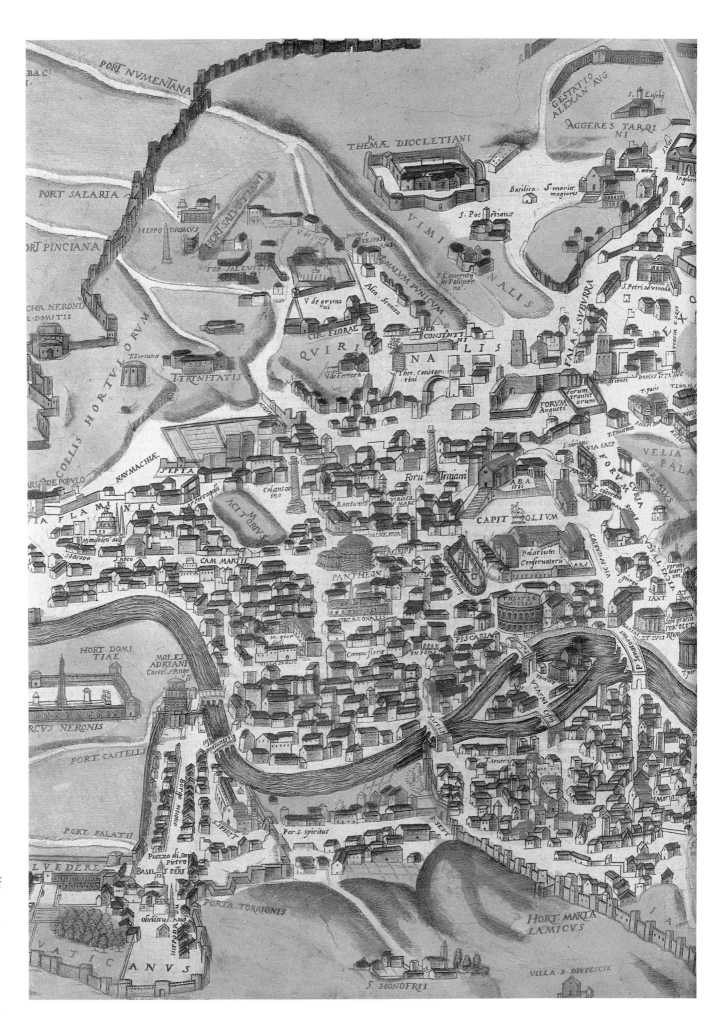

This bird's-eye view shows Constantinople, seat of Ottoman power from 1453, when the last Byzantine emperor was killed defending the great Christian city against the Turkish hordes of Mehmet Fatih (the Conqueror). Many buildings were damaged during the siege or had so long been neglected that they were in ruins, but a great period of restoration led to the building of the Suleymaniye mosque complex and the Topkapi Sarayi, the Imperial Palace.

The city was protected on two sides by water, the sea of Marmara and its extension into the

Yedikule Castle at the Marmara end of the city wall is part Byzantine and part Turkish. Embedded within its walls is the famous Golden Gate, the doors of which were formerly covered with gold plate. A Roman triumphal arch, it was erected outside the city of Constantine the Great in about 390.

The clearly marked Arsenal is in the wrong place; it was actually located on the Golden Horn. Visitors to the city were always much impressed by the arsenal which could hold 120 ships at a time. The Ottomans' great sea power depended on the arsenal, which built and maintained the ships that controlled the Mediterranean during most of the 16th century. But they suffered a great naval defeat at the hands of the Venetians at the Battle of Lepanto in 1571.

These portraits show the Ottoman sultans and their ancestor Osman, who, in the late 12th century, led a small group of nomadic Turks. His son, Orkan, is considered the first Ottoman sultan. This portrait, the fifth from the left, is of Mehmet Fatih, the conqueror of Constantinople; Murat III, sultan when the atlas was published, is at the far right.

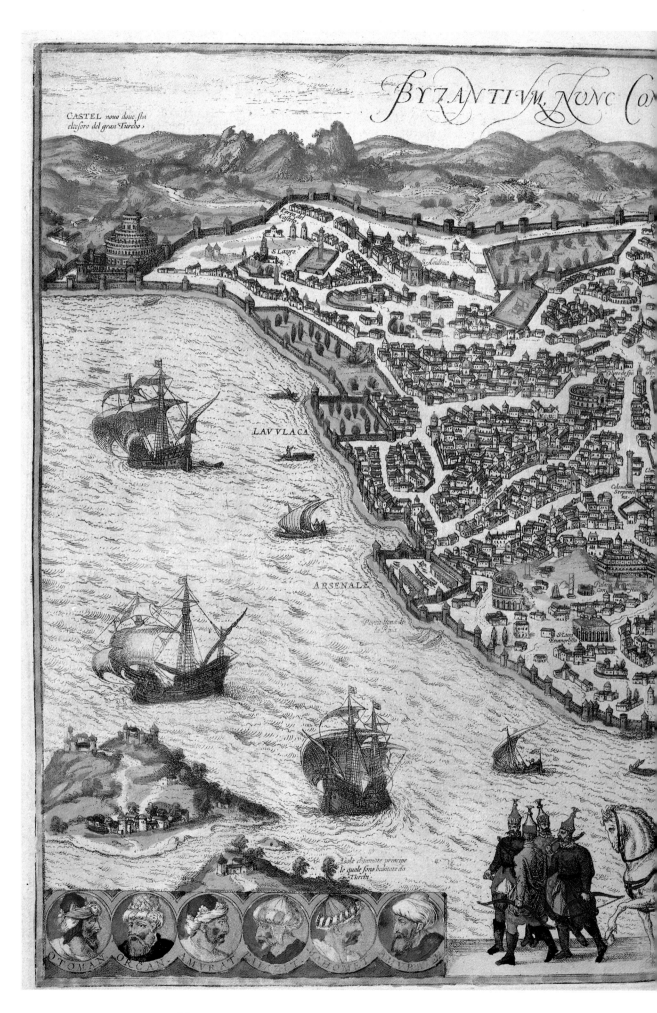

Golden Horn; on the right, the Bosporus leads into the Black Sea. On the north side of the Golden Horn lies the city of Galata, enclave of the Genoese merchants. Beyond, in the fields of Pera, the palatial ambassadorial residences and merchants' houses were yet to be built.

The greatest building within the walls is the church of Hagia Sophia (Holy Wisdom), just to the left of the Imperial Palace. The view is surprisingly accurate and even includes the ruins of the former hippodrome, which lies behind the arsenal.

The Palace of Blachernae was the imperial residence during the last centuries of Byzantine rule. The first edifice, built about 500, was much enlarged in the 11th and 12th centuries and its fabled and actual splendours probably fanned the desire of the Crusaders to possess the city, which they wrested from the Comneni dynasty in 1204 and ruled until 1261. When Byzantine rule was restored, a new dynasty, the Palaeologi, used the palace until the fall of Constantinople to the Ottomans.

The Suleymaniye mosque complex, although it is not the largest in the city, is the finest. It is a fitting monument to Sultan Suleyman the Magnificent and is the masterpiece of  its architect, Sinan. Construction began in 1550 and was completed by 1557; the complex shown here was finished a few years later. Only two of its four minarets are depicted, but the mosque truly dominates this view of the city, emphasizing its importance, which is undimished to the present day.

For 400 years the Topkapi Sarayi was the imperial residence and the centre of Ottoman power. The site was the acropolis of ancient Byzantium and was also chosen as a location for a palace but was later abandoned in favour of Blachernae by the Byzantine emperors. Mehmet Fahti built his palace on the high ground, laid out the slopes as an extensive park, and surrounded the area with a massive wall from the sea gate, protected by two large cannons, to the walls along the Golden Horn. This cannon gate, Topkapi, eventually gave its name to the whole complex.

A customary mix of Christian and mythical symbolism appears on the title page of Volume Five of the atlas, as well as pictures of exotic animals and two towering obelisks. At the bottom, a group of men in contemporary dress (including a Turk) sit in deep discussion.

Hamburg, a densely populated city in its early prime, is depicted here in a style lost for ever through the devastations of fire in 1842 and war a century later. The map shows an ideal site for a trading port—a settlement situated on the right bank of the River Elbe and within the basin of the Alster. The array of ships, both in number and variety, indicates Hamburg's importance as a centre of trade and the prominent position it held for centuries in the Hanseatic League. The dam on the Alster River, which created the city's lake, is featured at the centre right. The most important buildings are numbered on the map and annotated in the cartouche at the bottom of the plate.

In the early 1500s, the Portuguese set up trading posts in Asia and Africa to try to break the Muslims' monoply. This plate shows, *top*, views of Anaffa (Annaba) in Algeria, near the site of the Roman city of Hippo Regius, and Ruspina (Monastir) in Tunisia. In the centre is Diu, a small colony on the Gulf of Cambay, north of Bombay; below it is the much larger Goa on the spice-rich Malabar coast. Both of these featured prominently in early maps because of their importance in East–West trade.

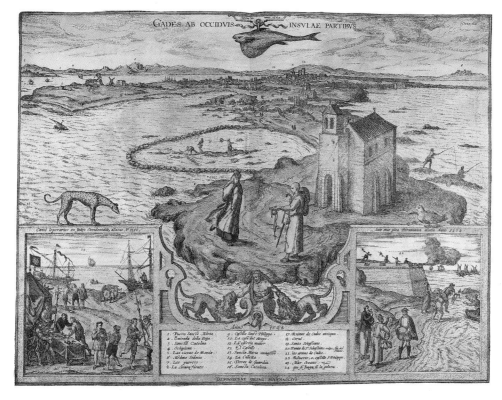

The city of Cadiz, the Gades of Roman times, is situated on a long narrow peninsula that juts out into the Bay of Cadiz. After the discovery of America, it became the home port for Spanish treasure ships and, as a result, the object of many raids by Barbary pirates. In 1587, soon after this view was drawn, an English squadron under Sir Francis Drake burned all the ships in the harbour.

47

GERARD DE JODE 1578

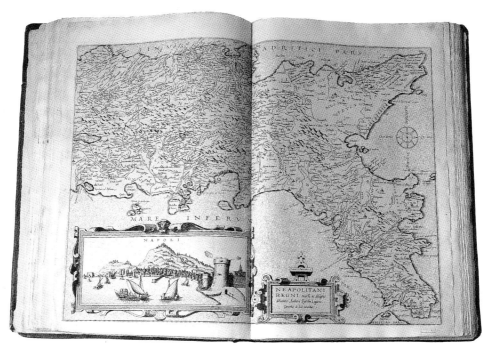

The Kingdom of Naples included all of southern Italy and with Sicily, the tip of which appears, formed the Two Kingdoms, which were united in 1504. The map shows the great sweep of the Bay, from the Isle of Ischia round to Capri.

The Kingdom of Naples, drawn by Pyrrho Ligorio

Jode used the word *speculum* (mirror) in the title of his work, and his map of southern Italy shows a delightful framed "reflection" of Naples in the late 16th century, with the old fort and busy harbour and Vesuvius towering over the city.

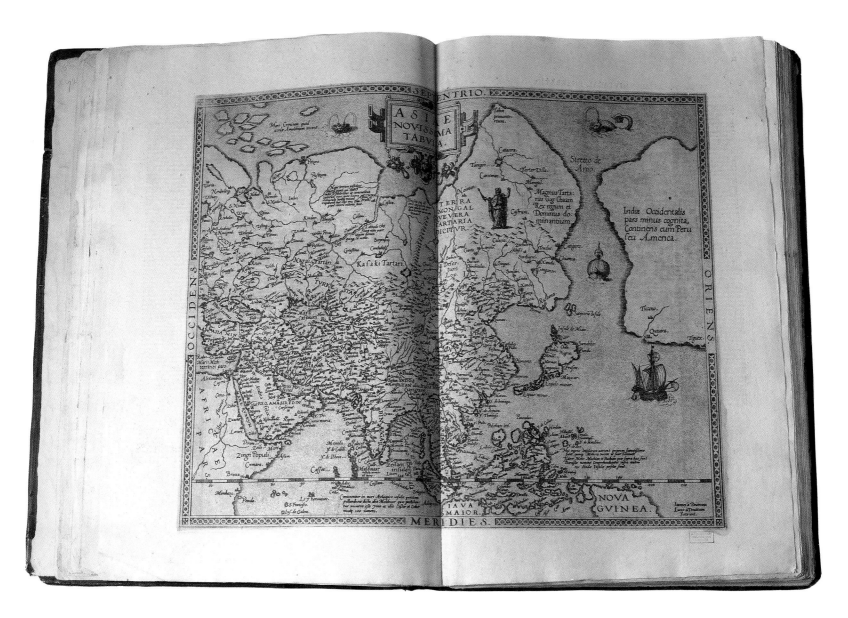

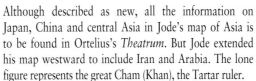

Although described as new, all the information on Japan, China and central Asia in Jode's map of Asia is to be found in Ortelius's *Theatrum*. But Jode extended his map westward to include Iran and Arabia. The lone figure represents the great Cham (Khan), the Tartar ruler.

MERCATOR, the dominant figure in cartography in the late sixteenth century, was quick to realize talent in others; for example, he complimented Ortelius on the "care" he had lavished on the *Theatrum*. But rivalry between the cartographer-publishers was not always so amiable; Ortelius and Gerard de Jode were on particularly bad terms.

De Jode's *Speculum orbis terrarum*, published in 1578, contains 27 general maps and 38 of Germany. Neither Ortelius nor de Jode mentioned the other in their lists of sources, although there were borrowings in both directions. Unlike the atlases of Mercator-Hondius, Ortelius and Blaeu, de Jode's *Speculum* ran to only two editions.

Cherubs representing the winds decorate Jode's maps. Here Mistral, a dry northwest wind, appears above Garbinus, blowing from the west.

Graecus, a northeasterly wind, is shown above Sirocco, a hot, oppressive wind from Africa that sweeps Europe in the spring.

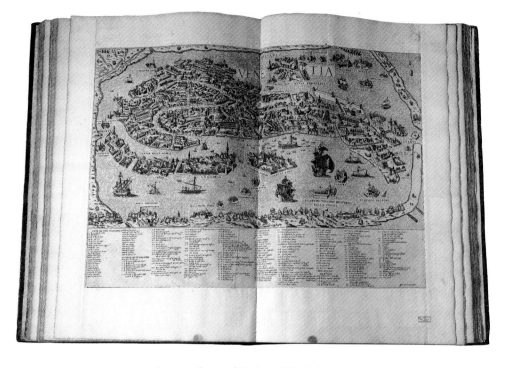

On the folding plate of the world map, *below*, much of the southern hemisphere is still occupied by a mythical landmass labelled Terra incognita: Ptolemy's Terra Australis.

The Bering Strait is not yet recognized and America and Asia are depicted as joined, while Japan, here called Cimpangu, appears only as a small island in the North Pacific.

Annotated map of Venice, within its lagoon

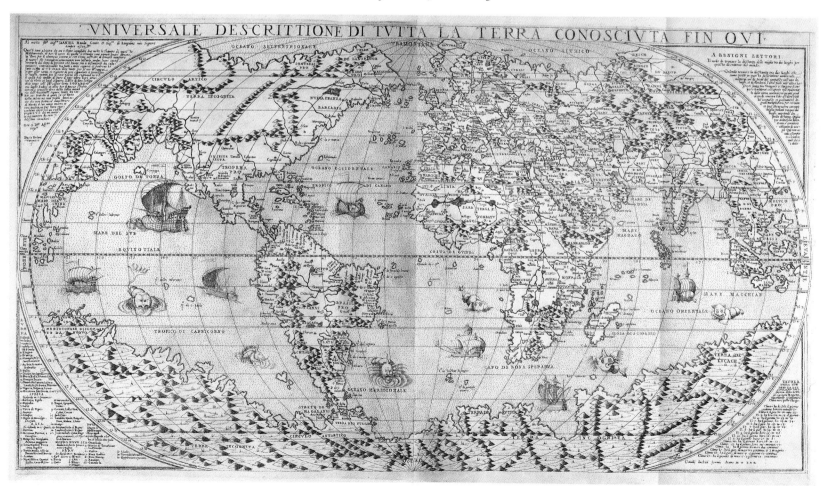

A view of the fortified city and harbour of La Rochelle on the west coast of France includes an inscription, "now occupied by heretics and rebels against his most Christian Majesty"—a reference to the Wars of Religion, fought intermittently from 1562 to 1598 between Catholics and Huguenots. La Rochelle, a Huguenot stronghold, successfully resisted six months of attacks by the Catholics in 1572–73. It is this siege that is commemorated here.

DURING THE sixteenth century, the grouping of maps within a binding became a regular practice in Italy. A notable collection was conceived by Antonio Lafreri, an engraver from Besançon who had settled in Rome in 1544. By 1553 he was publishing volumes of maps, each of which had a uniform engraved title page.

Virtually all surviving Lafreri atlases contain individual maps not found in the other atlases, and every atlas contains rare, occasionally unique, maps. Many include engraved views of cities and, less often, major events such as battles.

The view of Venice, *opposite*, is a fine example of Lafreri's detailed work. The city has changed little in the last 400 years and many of the buildings, identified by number at the foot of the map, are still recognizable. Mercantile activity and the many skilled engravers and printers who worked there made the city an important centre of map production.

The folding map of the Holy Land, *overleaf*, is presented so that west is at the bottom of the map. The title at the foot of the plate reads, Totius Terre Promissionis (the whole of the Promised Land). Important features include the Dead Sea, Jerusalem and the Sea of Galilee; on the far left, Damascus is marked as "The capital of Syria".

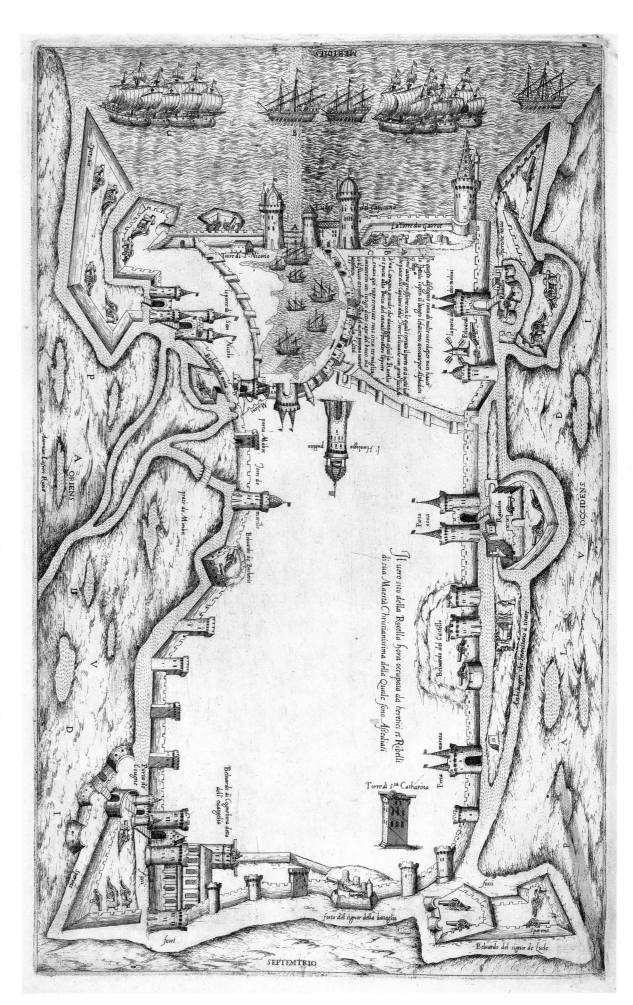

Sardinella Vrbs magna Chri, tianorum

Hic Abraham occidit Chodorlaomor regem elamitarum, et tres alios reges

Phiala fons

per plenus.

Gaualon Ciuitas

Mons Seir

Ramot galaad ciuitas in ea iehu vnctus fuit in regem israel.

amascus vrbs nobilis quæ et ram in Syria, abundans vino nguissimo, et lana præcipua. uus conditor fuit chore lius Esau.

Tota terra Orientalis a Mari Galileæ usq Cedar uocatur Traconitidis regio

DAMASCVS CAPVT SYRIAE

Sucta Vrbs

TABERNACVLA CEDAR

Nigra sum, sed formosa sicut tabernacula Cedar.

Ramoth Galaad. in ea fu

Ephrem. vt xps latuit

Introitus in emath

Sepulchrum Iob

Regio Traconilidis

In hoc loco fuit conuersio s. Pauli.

Mensa domini de septem panibus Math. 15 cap.

Cedar Vrbs et Regio 95 m. alitatium siue Saracenorum.

Hic Og rex basan cu oi exercitu suo occisus est a filijs isr el et sexaginta eius vrbes capte.

Edray nunc Adara

Gerasa Vrbs

Pella Vrbs

Iabes Galaad

Maspha in ea israel more suum vnicu

Hic xps expulit legi one a duobus demonia cis: que intrauit in porcos ex xps per missione.

Nemus Iabes

Masphæ Vr

Balnea calida

Ie Iacob transiuit Iordanem

Hic Stetit yhs in loco campestri

Corozaim

Hic yhs fecit Ser monem, in monte

Caphar naum

Mensa domini de qua pam ib hus.

Dimidia Tribus Manasse

Gadera Vrbs

MARE GALILEÆ SIVE TYBERIADIS siue stagnum Genezareth.

Balnea

Nota quod ista uallis ab utraq parte Jordanis uocatur Esdrelon.

SILVA seu uenatio Regum Damasci

Venatio Regia

Bethsalda

Hic domnus apparut apostolus piscantibus

Magdalum

Tyberias oppidu nunc Chennereth

Genezareth

Hic fuerunt S regis israel

Mons Gelboe

AQVAE MORON

Iordanis

Venatio Regionis

Buthuliam erat Iudich

Mons Hermon

Mons Hermon

is Suha.

Suha siue Turris Libani

Iord.

Hic cæsara vic tus est a barach

Cisterna in qua Ioseph fuit positus

Hic fratres ioseph pascebat greges

Dothaim

Hic Eliseus profeta percussit exercitu regis syrie cecitate

Tria tabernacula

Israel vrbs hac præcipu fuit iezabel mandato re gis israel.

Cæsarea philippi quæ paneas dicitur et belina deinde vocata est lais, et lesen, quam filij Dan accipientes vocauerut Dan siue Leasen.

DAN

Hic Iosue Dux israel pugnauit contra iabin regem Asor, Iobab regem madan, regem Semeron, et alios.

Neptalim Ciuitas

Hic Saladinus turcæ rex babilo niæ et damasci Solda nus vicit GALI xps LE A nor A

Saphet

Neptalim Tobiæ

Abel menla.

Tabor siue Ithabyrium mons

Ruma

Endor

Naim hic xps suscitauit filium viduæ

Philippus frater herodis tetrarcha ituree et trachoni tillis re gionum, cui in honorem tiberij Cæsaris Cæsaream philippi quæ nuc paneas dicitur appellauit, et est in prouincia phenicis.

In hoc Taberna culo occisus est Cisara a Jahal.

Chabul.

Hic madianitæ et alij occiduntur a gedeone.

Nazara

Suna Vrbs hic mansit heliseus pha in do mo mulieris sunamitis.

Aphec

Hic ac Israel pu contra S

Asor Vrbs munitissima.

Vallis Sommim iudicum. 4.

Naason Tobiæ

Tribus zabulon.

Saba

Messa

Asers ti. Georgy

Cana Ga lilea vbi Ierus fecit vinum

Sepheron

Capt diue virginis

Thoru Castru

Castru regis

Tribus ysachar

TRIBVS ASER Mons fortis.

Ista terra dicit capus magnus esdrelon: vel dicit campus maeeda vel planities galilee habet in logitudi ne viginti miliaria et duodecim in latitudine.

Antilibanus Mons.

CAYMOT

Mons Carmelus

MAGEDO

Suhala siue Suba

Tyrus siue sor phenicis Regionis metropolis

Iudin

Torrens Cison.

Rex huius vrbis occisus fuit a Iosue.

Putius aquar.

Hic helias occidit profetas Baal.

MONS MAGED

idon vrbs phenicis insignis, olim ananeorp terminus ad aquilone on longe a monte libano sita, et ostea regionis Iudeæ.

Regio Phæ nica

Hic Cananea occurrit Chro

Capa Helia

Hic mulier dixit beat

Sepulchrum Origenis

venter et porta tua.

Sandalium Castrum.

Ptolomais olim ephtimae autem dicitur acon Diuus Hier. in amos prof cap. p°

Gecmam Rex huius vrbis occisus fuit a Iosue

S. Marie

Cæsarea palestinæ quondam pyrgos, Sed ab herode rege iudeæ pulcrius, et contra vim maris, instructa: in honorem Cæsaris augusti, cesarea cognominata est.

Sarepta Si doniorum. Hic mulier vidua pas cebat heliam pro phetam.

Hieruptium

Spelonca Heliæ profete

Cesen.

Cæsarea palestinæ

SIDON Vrbs et Portus

Tyrus.

Casale R amperti

Ptolomais.

Siccamuon oppi

Castru peregrinor siue petra incisa

TOTIVS TERRE PROMISSIONIS A DAN VSQVE BERSABEE VERISSIMA ET AMPLISSIMA DESCRIPTIO.

Romæ apud Claudium Duchettum.

GERARDI MERCÆORIS ET I. HONDII.

ATLAS
ou
REPRESENTATION DU
MONDE VNIVERSEL, ET
DES PARTIES D'ICELUI,
FAICTE EN TABLES ET
DESCRIPTIONS TRES-
AMPLES, ET EXACTES:
Divisé en
DEUX TOMES.
Edition nouvelle.
Augmenté d'un Appendice de plusieurs
nouvelles Tables et Descriptions de
diverses Regions d'Allemaigne,
France, Pays Bas, Italie et de
l'une et l'autre Inde, le
tout mis en son ordre.

A
AMSTERDAM
chez Henry Hondius,
demeurant sur le Dam, a
l'enseigne du Chien vigilant.
Aº D. 1633.

CHAPTER THREE

The DOMINANCE of the DUTCH

THE NETHERLANDS in the fourteenth century were part of the dominions of the Dukes of Burgundy and in the early sixteenth they became united to Spain under the Holy Roman Emperor, Charles V. The struggle for freedom and religious independence began soon afterward and reached a climax in the reign of Philip II, who sought to introduce the Inquisition and to maintain a standing army in the Netherlands. William the Silent, Prince of Orange, the king's lieutenant in Holland, became the peoples' leader, and by 1581 the northern provinces had declared independence from Spain.

Because the Dutch had lived in a state of constant struggle with the sea for their very survival (Holland means the hollow land—indeed, it is almost a series of islands on a watery plain) they had become skilled mariners and were well able to take on the Spanish ships sent against them. In 1609, Philip's successor sued for a truce, which lasted for some 12 years, during which the Dutch burghers recouped their financial losses by trade. After the armistice broke down, hostilities were resumed and fighting continued until 1648 when, in the Treaty of Westphalia which ended the Thirty Years' War, Spain finally recognized the independence of the Low Countries.

The sumptuous title page, *left*, adorns Volume One of the Mercator-Hondius atlas; the tooled vellum binding of Blaeu's atlas is shown above.

It was the Dutch who first challenged the monopoly of the Portuguese and Spanish in Eastern waters. Throughout the struggle with Spain they increased their domination to such an extent that they had, by the middle of the seventeenth century, virtually supplanted Spain and Portugal as the carriers of oriental cargoes, and Dutch power and trade increased substantially with the founding of the Dutch East India Company in 1602.

The English, too, had established an East India Company, in 1600, and were flexing their trading muscles. The Navigation Acts of Oliver Cromwell, which sought to ensure that imports were carried only in English ships, in 1652-54 provoked war between these two fledgling maritime powers. In 1662-65 another conflict saw the Dutch Admiral Michiel de Ruyter enter the Medway and threaten the safety of London itself.

In 1672, however, William of Orange became stadtholder and five years later married Mary, daughter of James, Duke of York and later James II of England. In 1688 William was invited to become joint monarch with Mary and reigned as William III of England until his death in 1702. During this period and later the Dutch and English fought side by side. They inflicted many defeats on Louis XIV of

France until the Treaty of Utrecht in 1713 concluded the war, effectively ending the era of Holland's greatness and marking the rise of Britain as a world power.

The Netherlands became the home of Europe's most brilliant cartographers and hydrographers at about the time the Dutch East India Company came into existence in 1602, mainly because the Company established its own cartographic department and appointed appropriate personnel. The cartographers were assisted by the use of the Mercator Projection, which allowed flat maps to be made on which sailors could plot a straight course across a spherical globe.

The Spanish coat of arms, with the two black eagles of the Habsburgs, is used to ornament a map of Spain in Jansson's atlas of 1656.

Mercator was born of German parents in 1512, near Antwerp. He went to the university of Louvain, graduating in geography, geometry and astronomy. Shortly afterward, he established a business making globes, astrolabes and other scientific instruments, and later turned his hand to engraving maps and charts. In 1522, he moved to Duisburg and began work on his famous world map of 1569, which accurately depicted the coasts of Central and South America and included a more accurate outline of Asia, including southeast Asia, but perpetuated the Greek concept of a vast southern continent, Terra Australis.

The greatest feature of the map, however, was the cylindrical projection, in which all the meridians are straight lines perpendicular to the Equator, while the parallels are straight lines parallel with it. To achieve his purpose on paper, Mercator made the spacing of the lines of latitude progressively larger away from the Equator, in the same proportion as the spreading of the meridians. It is thus clear that places far to the north or south appear somewhat distorted on a map based on this projection. For instance, Greenland appears about the same size as South America, which is about nine times larger. The full significance of the map and its projection was not recognized for almost 100 years. The reintroduction, more accurately it must be said, of the Greek lines of latitude and longitude was another important feature, but one that is often overlooked.

Mercator's large map of Europe, produced in 1554, was also a considerable improvement on earlier maps based on Ptolemy, for the length of the Mediterranean Sea was reduced from 62 to 53 degrees (still 9 degrees in excess), and this was generally accepted for over a century, despite the fact that practical navigators probably realized that it was wrong. The latitudes covering Europe on the map are generally accurate, although to the north and east small errors do occur. A second edition incorporated information about the White Sea, Moscow and the interior of Russia gained by English explorers, which demonstrates the piecemeal way in which maps of the period were compiled.

Mercator regarded his world map as part of a publishing plan that began with an edition of Ptolemy in 1578 and culminated in the complete work, with a general title page, in 1595, the year of his death. Mercator was the first to use the word "atlas" as a term to describe a collection of maps. For several reasons, this work was not a great commercial success until Jodocus

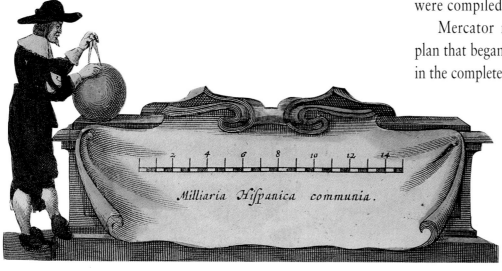

In another corner of Jansson's map of Spain, a figure dressed in Dutch-style clothes measures the distance on a globe with dividers; the key below indicates the Spanish scale of distances used on the map.

Hondius purchased the plates from Mercator's heirs and added further maps to make his edition more comprehensive. The first edition of the Mercator-Hondius atlas appeared in 1606 and it ran to some 30 editions before 1640. But it was soon challenged by the monumental work of Willem Blaeu, which appeared in the 1630s.

A close contemporary of Mercator was Abraham Ortelius—the Latinized form of Ortell—again the son of German parents who had settled in Antwerp. Ortelius was born in 1527 and became a noted mathematician before turning almost exclusively to geography and cartography. In 1570 the renowned Plantin press published his *Theatrum orbis terrarum*, which is widely regarded as the first immediate commercial success, since it met, in a suitable format, a demand for knowledge of the new lands being explored.

Map makers used dedicatory coats of arms, such as this one in the Mercator-Hondius atlas, to solicit or repay patronage.

In 1575 Ortelius was appointed geographer to Philip II of Spain, a post which gave him unrivalled access to the accumulated knowledge of the Portuguese and Spanish explorers. Ortelius sought to combine all their cartographic and geographic material with that from other sources, such as the information about Russia published by Anthony Jenkinson, so as to offer, in his definitive edition of the atlas in 1595, a faithful picture of the known world.

One whole family of cartographers, instrument makers and publishers which dominated the seventeenth century was that established by Willem Janszoon Blaeu (1571–1638). Born at Alkmaar, he studied under the great astronomer Tycho Brahe and set up on his own account in 1596 at Amsterdam. He was making globes for about the first 10 years, but then began to produce maps of Holland and Spain, as well as a sea atlas in 1623. In 1629 he bought some plates from Hondius and issued his first atlas in 1630. His more famous son, Joan (1596–1673), collaborated with his father and brother in the production and expansion of the land atlas until the *Grand Atlas* of 1662 comprised some 11 volumes and a sea atlas.

Amsterdam soon became the European centre for the making of maps, globes, charts and navigational instruments. This work was dominated by the three family firms of Hondius, Jansson and Blaeu. Of these cartographic publishing families, the Blaeus were the most famous, and because of their official positions within the Dutch East India Company, one would expect the maps to show all the latest information, but they often lack details which were well known to navigators. Perhaps this was due to official reluctance to reveal sensitive or useful information, but the Blaeus were probably also mindful of the fact that the purchasers of these sumptuous volumes were unlikely to require absolute accuracy.

This cartouche appears in Doncker's atlas, on a map of the African coast from Angola to the Cape of Good Hope. The black figures, garbed in Arab style costume, hold nautical instruments.

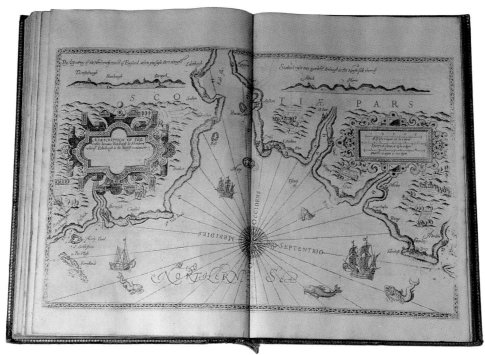

The Scottish coastline from Bamburgh to Aberdeen, with the Firth of Forth

BEFORE THE START of the fifteenth century, European navigators relied on books of sailing directions. Known in England as "Rutters", a corruption of the French "Routiers de mer", they provided information about tides, currents, rocks, sandbanks and other hazards. Later, charts supplemented, then superseded the books and, as ocean travel became more common, parallels of latitude were inserted; the problem of the convergence of the meridians at the poles was overcome only when Mercator perfected his projection.

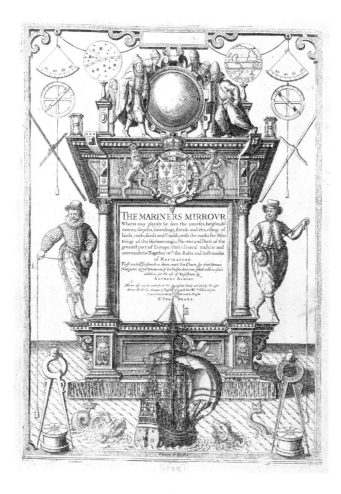

The flamboyant, almost theatrical, title page of the English edition of Waghenaer's work, which was translated as "The Mariners Mirrour", displays a superb range of nautical instruments, including the cross-staff, astrolabe, dividers and plumb lines, as well as a fine representation of a ship of the period and fanciful marine monsters.

Books and manuscripts were produced throughout the 1500s, charting coastlines and telling pilots how to fix the latitude of landmarks and take bearings from them. In 1584 the chart maker and pilot, Lucas Janszoon Waghenaer, published *Die Spieghel der Zeevaeri*.

Printed by Christopher Plantin from plates engraved by Johannes van Doetecum, it covered the main ports and important coastlines of western Europe from Norway to Portugal and was the first to contain charts and sailing directions in one book.

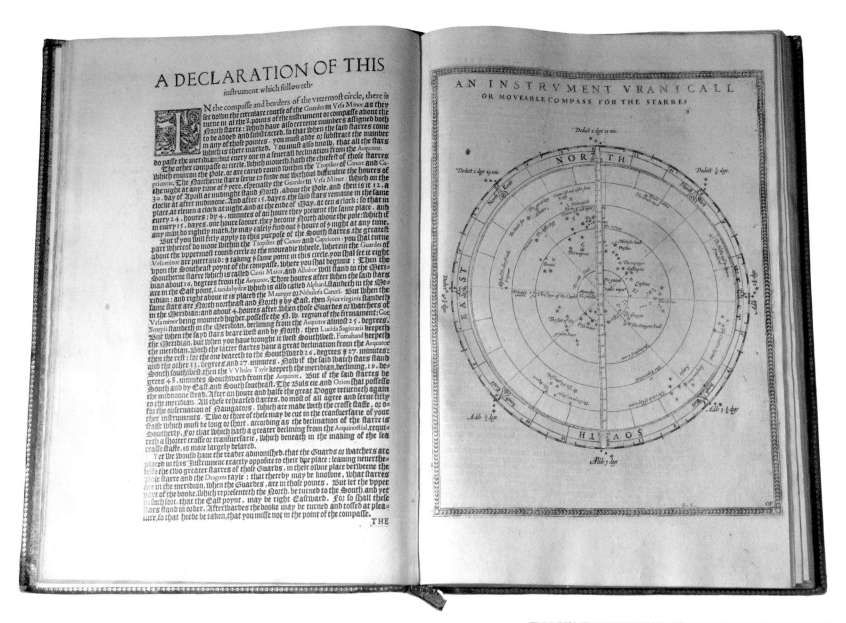

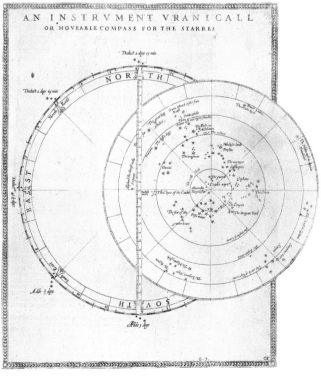

The "moveable compass for the starres" comprises an outer circle, showing eight points of the compass, and an inner disc which identifies the main stars of the northern hemisphere and can be rotated, *below*, allowing the mariner to "finde out without difficultie the houres of the night at any time of ye yere". East and west are reversed; the text instructs the user to rotate the book through 180 degrees so that they are in their correct positions, when he can move the inner disc until the stars correspond exactly with those above him.

Map makers often filled the land and seas with pictures of exotic creatures, sometimes entirely fanciful and at others, as with this whale, quite closely observed.

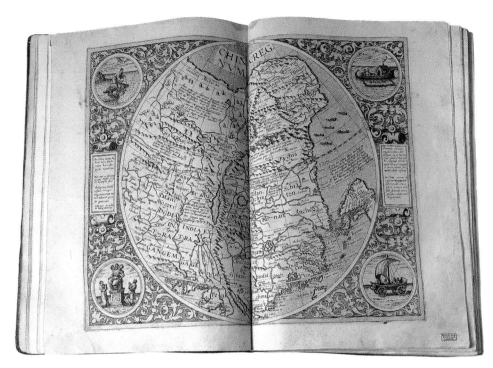

Map showing China and part of Japan

The map of China, *above*, covers the area from the Bay of Bengal to the Pacific and includes the Great Wall. Beyond this is the region designated Tartary—present-day Mongolia. Mongol yurts are shown, and the inscriptions reveal that the region was ruled by the Chinese. Japan has been incorrectly drawn to show land extending to the southeast, rather than the north.

The four vignettes give curious information: top left shows that the Chinese used birds to catch fish and, top right, that some families lived on boats and used floating fences to keep aquatic livestock corralled; at bottom right is a wind-propelled wheeled vehicle. The last roundel pictures a *ju-ichi-men kannon*, or 11-headed *boddhisatva*, worshipped in Japan.

A map of the Straits of Anian (the Bering Strait) separating Asia from America, *left*, shows an open sea passage north of the American landmass. It was this, shown on many maps of the period, that prompted European explorers to seek a northwest passage.

The plate, *right*, covers the Solomon Islands and part of New Guinea, south of which is a land inhabited by dragons, lions and snakes, referred to in the caption as "terra Australe".

GERARD DE JODE'S *Speculum orbis terrarum* was published in two volumes at Antwerp in 1593 as an enlarged version of the atlas first produced in 1578. The first part includes an introduction and essential information regarding the mathematical bases of cartography.

This is followed by maps of the world and spheres showing the north and south polar regions; the latter is still named Terra Australis incognita. There are also general maps of South America,

Africa, Barbary (including much of north Africa), and Asia from the Red Sea to the Pacific Islands.

The remainder of the first part is devoted to maps of Europe, as is the second volume, which comprises detailed maps of both Germany and the Low Countries.

This 1593 edition was published two years after Gerard's death by his widow and his son Cornelis, but neither edition of Jode's atlas achieved the popularity of Ortelius's *Theatrum* of 1570.

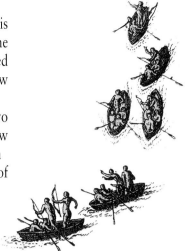

In this bird's-eye view of North America, the east coast and the northern regions, including the St Lawrence River, Newfoundland and Labrador, are prominent, while a large part of the country is designated New France. The cartouche shows native inhabitants of Virginia, including one hunter with a bow as tall as himself.

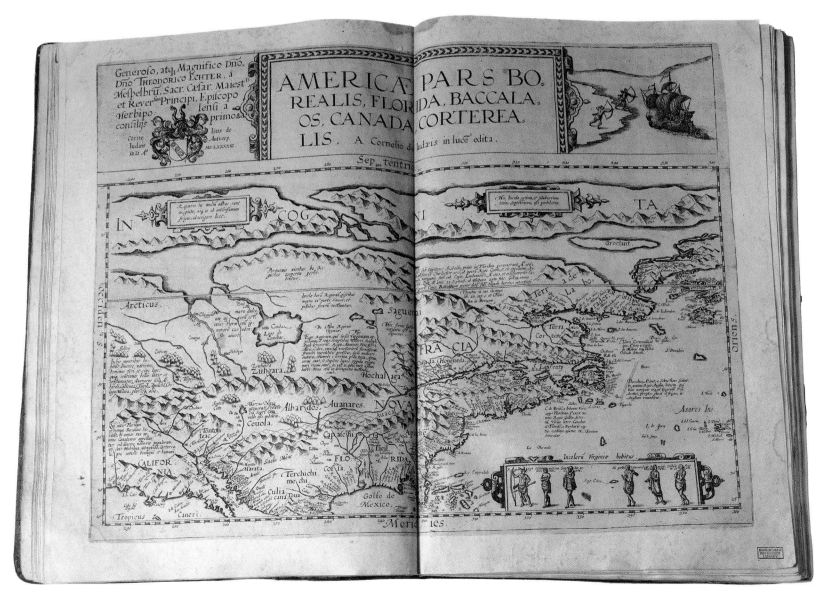

Title page of Linschoten's *Discours of Voyages*, engraved by William Rogers of London

The market place in Goa shows European-style houses and Portuguese merchants conducting their business protected from the sun by parasols.

J AN HUYGEN VAN LINSCHOTEN (*c*1563–1611), a Dutch traveller and writer, was born in Harlem. The text of his *Voyages into the East and West Indies* was translated into English in 1597, but the captions on the plates remained in Latin and Dutch. Linschoten travelled to Goa in 1583, where he was employed by the archbishop. His account of his voyages was of such value to navigators that it became almost a seamen's manual.

A lady walks with two Indian maids, and the pet dogs of the Portuguese play in the street. Some Indians wear a cross to show they are Christians.

Cochini Rex elephante vectus, cum procerum comtu, quos Nairos vocant.

Die Coninck van Cochin op een elephant geseeten verselschapt met sijn edelen diemen Nairos noemt

64 en 65

Linschoten's book, *Voyages into the East and West Indies*, provided detailed surveys of trade routes and translations of Portuguese and Spanish documents on geography, statistics and navigation. By revealing the weakness of Portuguese administration in the East Indies, it acted as a stimulus to the Dutch and English in their attempt to dominate the region; in 1663 the Dutch took Cochin; they were displaced by the British in 1795.

Most of the plates show scenes of Goa and its environs, based on the author's drawings and engraved by both Dutch and English craftsmen.

The King of Cochin is accompanied by a retinue of nobles called the Nairos. Cochin was an unimportant fishing village until water separated the area from the mainland, turning the harbour, founded in 1500 by the Portuguese, into one of the most serene ports on India's west coast.

Four typical Indian plants are shown in their natural setting, *left*. The caption states that the Indians eat the fruit and drink the milk of the coconut and use fibre from its trunk for making ships' ropes. The Indian fig gives fruit all year round. The seeds of the Areca palm (back right) are chewed with a wad of leaves from *Piper betle* as a purgative. Around its stem twines *Piperus frutex (Piper nigra)*, which yields peppercorns.

Linschoten's portrait was made in 1595, when he was aged 32. The likeness is surrounded by four insets. At the top appear Goa and Mozambique, on Mozambique Island. Both had been Portuguese strongholds since the early 16th century. The latter is dominated by the three forts of San Antonio, Santo Domingo and Santa Gabriela.

At the bottom are two different views of St Helena, which, although only a small, isolated island in the South Atlantic, was of great importance to sailors as a watering place. It was discovered by the Portuguese navigator João da Nova in 1502, then occupied by the Dutch, who in 1651 were succeeded by a garrison of the British East India Company.

Fresh meat, as well as "salubrious drinking water", was available on St Helena, as this view, *below*, showing hunters with guns and spears chasing goats testifies. Lying off the island is a group of named Portuguese ships—probably returning from the East. The plate is dedicated to Philip, Edward and Octavian Fugger, members of the famous German banking family, whose arms appear.

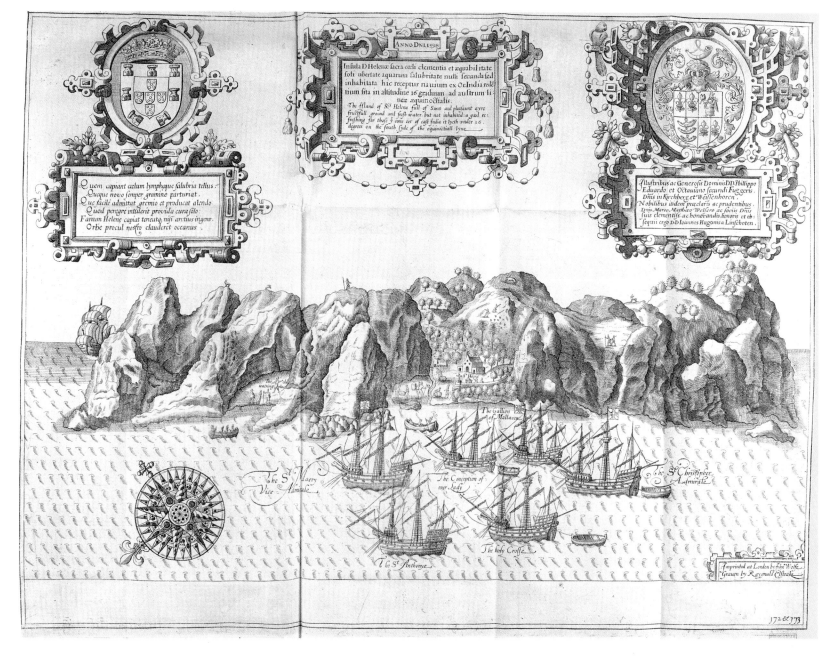

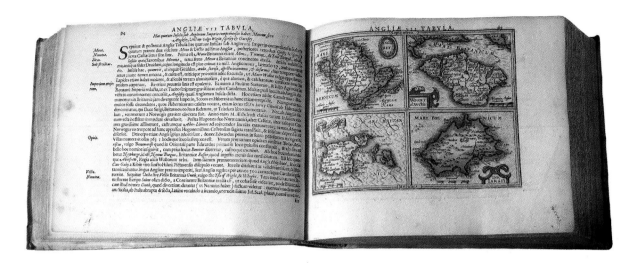

Plate showing islands off the English coast

T HE *ATLAS MINOR* of Gerardus Mercator was first published by Jodocus Hondius in 1607 and was often reprinted. Unlike the splendid two-volume edition of the Mercator-Hondius atlas, intended for the libraries of the wealthy, this small single-volume version was produced for the less affluent, emerging middle classes. Each plate was still hand coloured, but in a restricted palette; the typography, too, was less elegant.

The plates covering the British Isles include one that shows small islands off the English coast, *above*. It is meticulously annotated, and many of the towns that can be seen still exist today, among them St Hilary (St Helier) on St Aubyn's Bay in Jersey, St Peter Port on Guernsey, Newport on the Isle of Wight, and Holyhead on Anglesea—from which the shortest sea crossing can be made to northern Ireland.

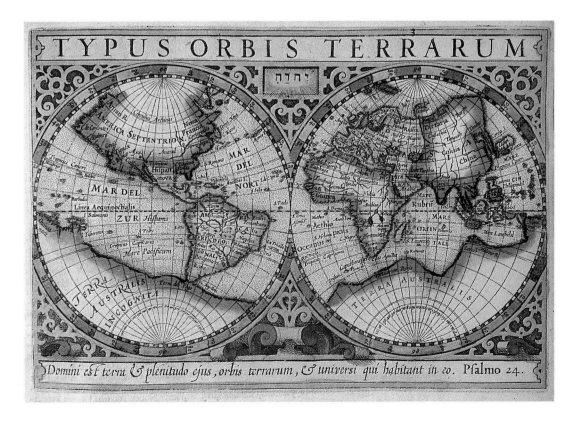

The world map, in two spheres covering the eastern and western hemispheres, is drawn on the Mercator projection and, as a result, exaggerates the unknown landmasses in the polar regions. The verse from Psalm 24 at the bottom of the plate reads: "The Earth is the Lord's and the fulness thereof; the world and they that dwell therein."

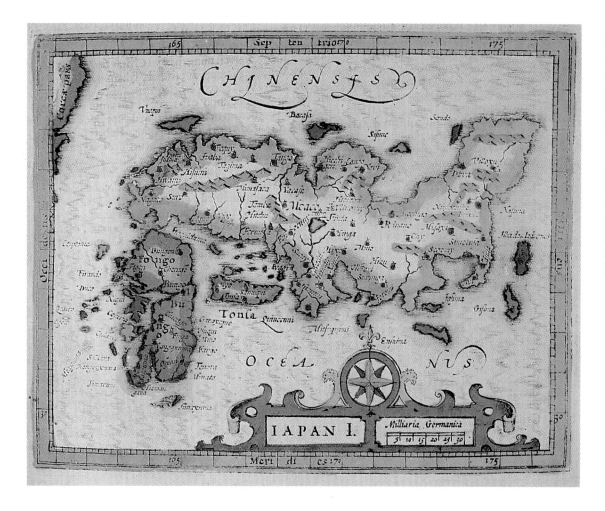

European merchants who traded with Japan visited only the island of Kyushu, where they obtained porcelain and lacquerware. Kyushu is set at right angles to the island of Honshu, giving the country an entirely wrong shape, and the map excludes the northern island of Hokkaido, which was also ignored by the Japanese, since its inhabitants, the Ainu, were not regarded as truly Japanese.

The accompanying text tells of the magnificent palace of the Taico (ruler—the origin of the word tycoon), with a hall comprising 1,000 tatami mats. Rooms were measured in these mats, each 6 x 3ft (2 x 1m) in size; a good-sized room would be about six mats. The text also mentions that a Jesuit seminary had been founded where Japanese and Portuguese learned each other's languages.

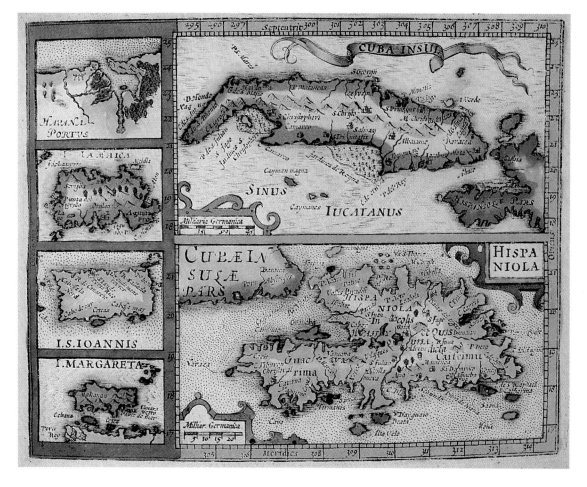

This plate includes Cuba and Hispaniola, with insets of Havana and the islands of Jamaica, St John and Margareta. Protected by the castle, Spanish treasure galleons could assemble in the safe harbour of Havana, founded in 1515, for their return to Europe and their home port of Cadiz.

The text printed on the page opposite the maps gives information about the vegetation, climate, animals and customs of the islands. "Sugar is found here and various kinds of fruit and herbs, grain and spices, including cassia, aloes, cinnamon and ginger...The native inhabitants are benign and content, they often walk about without clothes and only rarely put on tunics ..." made, it seems, from a type of bark cloth.

As with the two-volume Mercator-Hondius atlas, the title page shows the Titan, Atlas, one of the 12 children of Uranus and Ge—all of gigantic strength and size—supporting the world on his shoulders. In the Greek legend the Titan was condemned to support the sky for eternity to prevent its falling on the earth, but the legend became corrupted, and he usually appears holding up the globe.

The title page indicates that Jodocus Hondius contributed to the atlas and that Jan Jansson was involved in its publication.

The map of Paradise includes a vignette of Adam and Eve partaking of the forbidden fruit, while in the background they are shown being shepherded out of the garden by an angel. The Garden of Eden is confidently placed east of the Dead Sea, close to Petra.

The map also shows Chaldea, lying between the southern stretches of the Tigris and Euphrates rivers, and the cities of Ur, Sodom and Gomorrah. Other cities, including Damascus, appear as do many smaller settlements, and on the coast of the Mediterranean the cities of Antioch, Tripoli, Tyre and Sidon are marked.

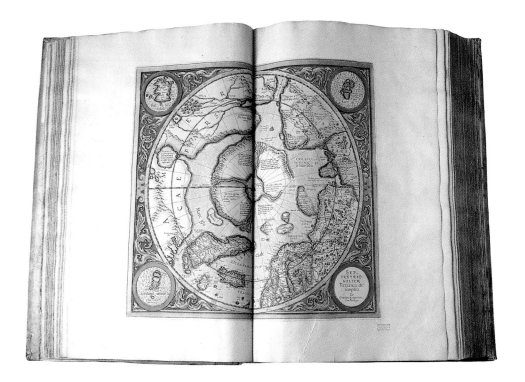

Map of the Arctic Regions, first published in 1569

This detail shows a Dutchman in contemporary dress beside East India House, headquarters of the Dutch East India Company. Built in 1606, it was extended soon after this view was made.

GERARDUS MERCATOR'S fame rests largely on his now extremely rare sheet map of the world (1569) in which he first used his celebrated projection. This allowed sailors to steer a long course without repeated adjustments of compass readings. The map contained an inset, which was repeated in the Mercator-Hondius edition of 1633, *above*, to cover the Arctic, since its depiction on this projection would have produced far too great an exaggeration of the landmass.

Open water shown around the pole gave hope to those who believed the East might be reached by a northeast or northwest passage; some even hoped to be able to pass by the "high black rock" of the North Pole. One of the magnetic poles appears as a rock within the straits between Asia and America, the other at geographical north. Inset are the Faroes, Shetland and an imaginary Frislant.

The man dressed in the "old style" stands next to a depiction of West India House, the headquarters of the West India Company. The building, which was erected in 1623, survives to the present day.

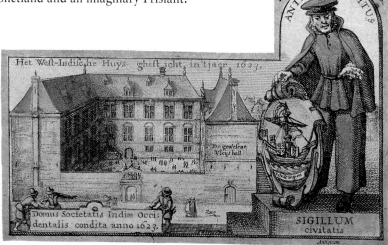

An aerial view of Amsterdam shows many buildings and landmarks that still exist. The Amstel River and the canal system remain the same, but the docks, an arm of the Zuiderzee, were greatly altered in 1872, so the inset view showing the city from the inland sea can no longer be seen. To the left of the plan is an outline of a proposed extension to the city's fortifications, which was later carried out. Windmills, marked all along the ramparts, were essential to work the pumps that prevented Amsterdam from flooding.

The mass of ships crowding the docks and tied up at the pontoons is clear evidence of the position of the Dutch as the greatest trading nation at the time and of Amsterdam as Europe's greatest port. The Dutch East India Company held the monopoly of trade with all countries east of the Cape of Good Hope and the exclusive right to sail through the Straits of Magellan, while the West India Company, established in 1621, controlled trade with the Americas and administered the New Netherlands, where New Amsterdam (later New York) was founded.

An extensive numbered list of the principal buildings appears on either side of the depiction of Amsterdam, *above*. Its comprehensive nature indicates the growing wealth and prosperity of the city and, indeed, is almost a celebration of the rule of the city burghers.

In addition to the important headquarters buildings of the East and West India companies shown in detail, *opposite*, there are two further insets. Top left is the Stadshuis, the old town hall, which was destroyed by fire in 1651; a new one was completed in 1655. At top right is a depiction of the Burse, where merchants used to meet to transact their business.

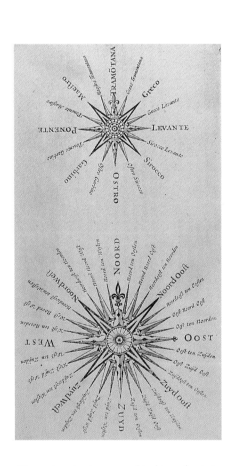

Illustrations in the geographical introduction to the atlas indicate, *above*, the winds that blow from the different points of the compass in the Mediterranean region; the sphere below names the points of the compass and details their divisions.

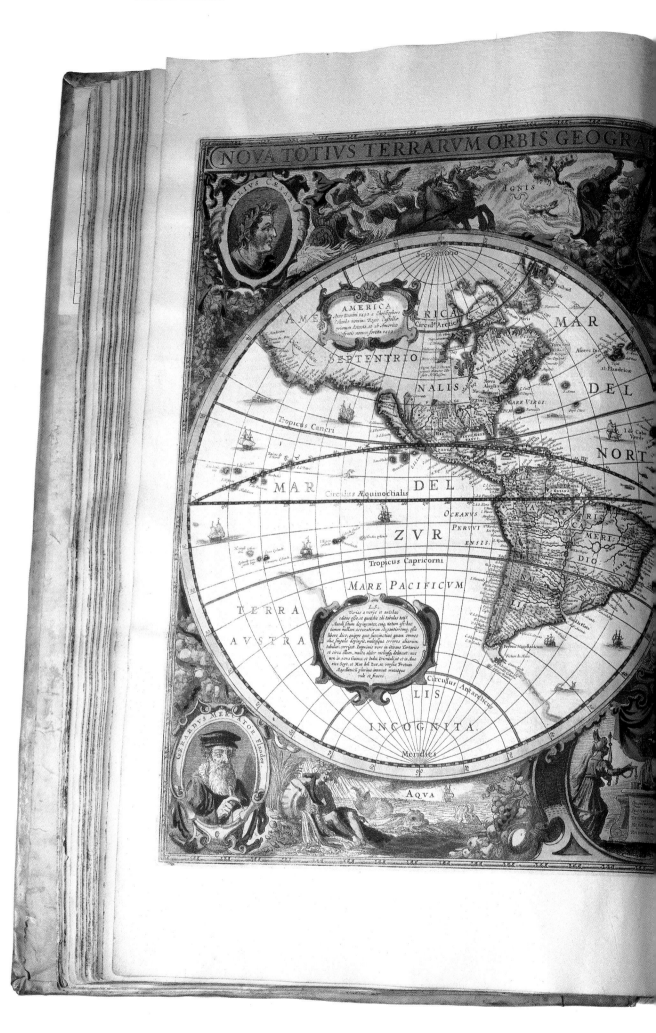

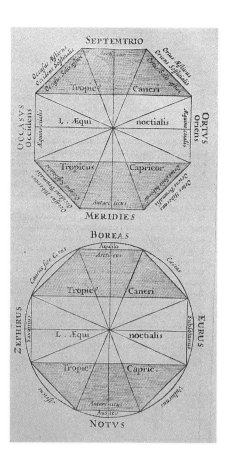

Here, the illustrations show the "four quarters" of the world. At the top, the world is divided at the Equator, with the tropics either side of it and the arctic regions at the extremities. The central meridian is also marked. The text accompanying the bottom sphere, indicating the prevailing winds, includes an elegy by the Roman poet Ovid in which he celebrates them.

The map of the world has at each corner a cameo portrait depicting the geographer Ptolemy, Hondius and Mercator, the begetters of the atlas, and, somewhat incongrously, Julius Caesar. Between these are scenes of the four elements as defined by the Greek philosopher Empedocles: fire, air, earth and water. The cartouche at the bottom appears to show personifications of the four continents, with Europe receiving homage from Asia, America and Africa.

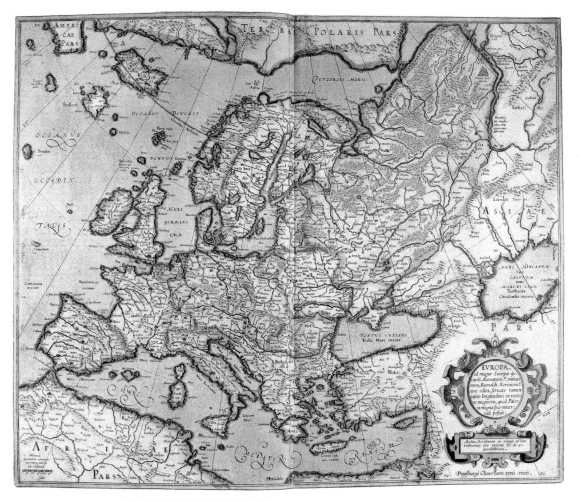

Mercator's map of Europe shows a clear water passage to the north of the landmass: the northeast sea passage that many explorers hoped would lead them to the East. They did not yet appreciate the icy conditions that prevail for most of the year nor the vast ice-cap of the polar region. The existence of a great continent north of Asia (here named Terrae Polaris Pars) was disproved by the Swedish explorer Nils Nordenskjöld only in 1878–9.

Iceland, Greenland and the Faroes are marked, but also a mythical Frislant Island. What is probably Baffin Island, discovered in 1615–16 by the English explorer William Baffin, is labelled "Part of America".

The title page of Volume Two, *right*, is an elaborate confection of classical motifs and allegorical symbolism. The classical façade is surmounted by an Indian woman with a feather head-dress (Mexicana) and a black woman (Africa) leaning against a crocodile and holding a parasol. Other niches contain figures representing Europe, Asia, Peru and Magalanica—the landmass supposed to exist south of Tierra del Fuego.

The map of Asia identifies many of the important trading stations used by Europeans at this period, among them Hormuz, Goa and Malacca. Lesser ports, such as Rasht and Amul in Persia (Iran), are also identified. Sri Lanka bears its Arabic name, Zeilam, from which the English Ceylon derived.

Many of the coastlines were much travelled and thus well drawn, but others are less accurately shown. The mythical Terra Australis Pars, with the entrance to the Gulf of Carpentaria separating the two areas of land, is beginning to assume a shape recognizable as Australia. Once again, an open-sea northeast passage appears.

The title cartouche records this as a map of America, or New India, the land early explorers sought. The other insets comprise maps of the Gulf of Mexico, Cuba with Jamaica to the south, and Haiti (Hispaniola).

The shape of both north and south continents is inaccurate: both are too wide, probably because of the Mercator projection used. The southern half of the sphere is dominated by the immense "Not yet known" landmass of Terra Australis, which Mercator accepted as existing. Tierra del Fuego, separated from South America by the Straits of Magellan, is shown connected to this continent. The Great Lakes of North America are sited too far north and the St Lawrence River is too long; Hudson Bay, by contrast, is too small.

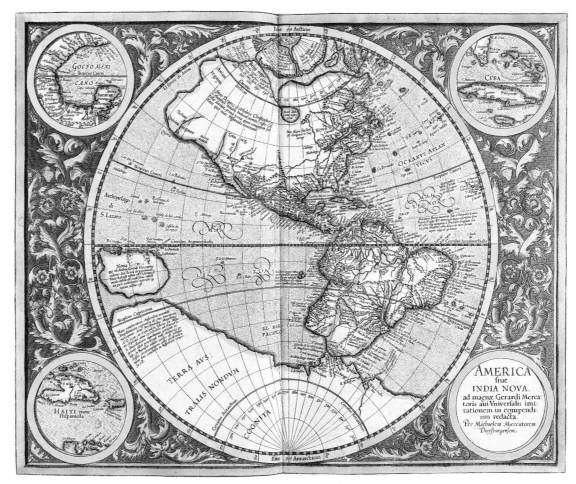

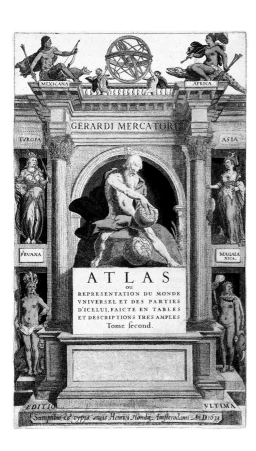

More than 100 years of exploration and trade gave Europeans a detailed knowledge of Africa's coastline, but little understanding of its interior. Thus, while both the Congo and Nile rivers appear, the latter extends too far southward. Ethiopia and Tigre province occupy excessive areas of central and southern Africa and the north coast. Although differently spelt, the great trading centres of Timbuktu and Kano on the caravan routes across the Sahara are accurately marked, as are the old kingdoms and regions of Senegal, the Gambia and Mozambique.

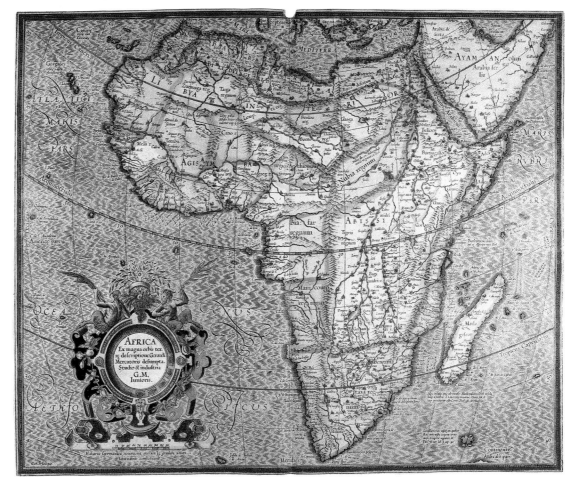

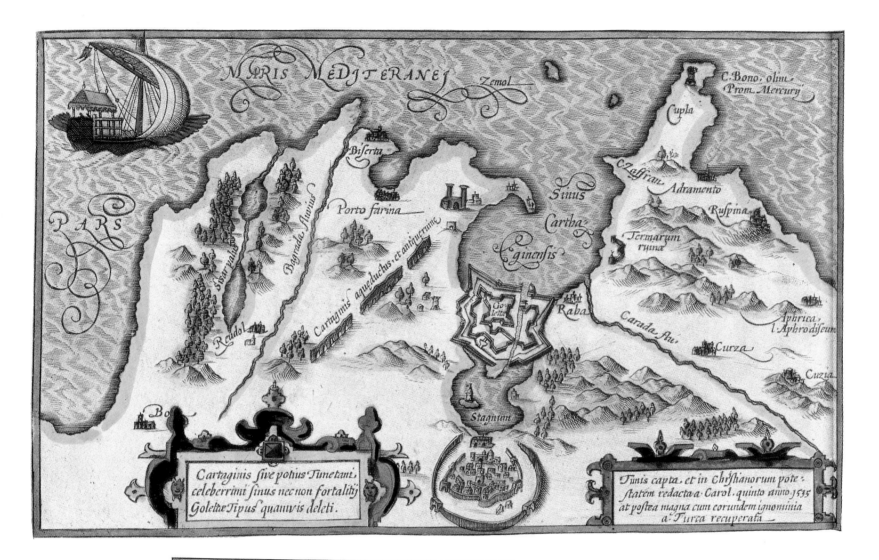

Among the insets on the map of the Barbary coast is a depiction of Tunis, set at the end of a shallow lake and linked by a canal to the Gulf of Tunis. This was the site of ancient Carthage, razed by the Romans in 146 BC, and the ruins are shown of the great Roman aqueduct, some 87 mls/140 km long, that brought water from the mountains inland.

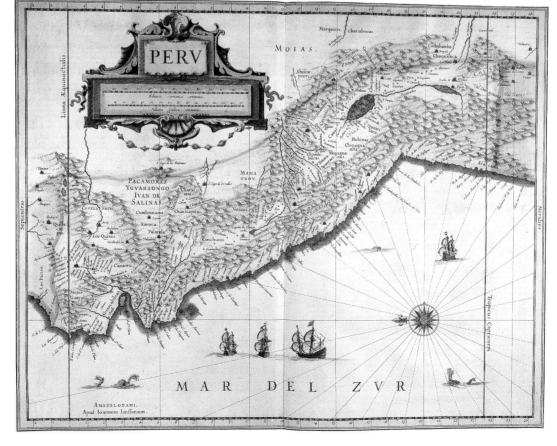

The map of Peru, compiled by Jansson, includes Ecuador and part of northern Chile, the coastline of which is distorted.

The map is orientated so that north is on the left-hand side, south on the right, and the Equator and Tropic of Capricorn are drawn in vertically. The Andes mountains and lakes Titicaca and Poopo are clearly shown.

GERARDUS MERCATOR NATUS
RUPELMUNDÆ.III NON.MARTII ANNO
CIƆIƆXII:VIXIT ANN.LXXXII.M.VIII.D.
XXVI:DENATUS IV NON.DECEMBRIS
ANNO CIƆIƆXCIV.

IUDOCUS HONDIUS NATUS IN
PAGO FLANDRIÆ DICTO WACKENE XVI
KALEND.NOVEMBRIS ANNO CIƆIƆLXIII:
VIXIT ANN.XLVII.M.VII.D.XXIX:DENAT
US XIV KAL.MARTII ANNO CIƆIƆCXII.

The celebrated double portrait of Gerardus Mercator (left) and Jodocus Hondius first appeared in the Mercator-Hondius atlas of 1612. Mercator died in 1594, and his son Rumold made many additions to the atlas. On his death in 1602, the plates were bought by Jodocus Hondius, who added further maps. In the year this collaborative atlas was published, Jodocus died, but his son Henricus and son-in-law, Jan Jansson, continued to add to the collection. The collaboration between the heirs of Mercator and Hondius was a very profitable one.

EPITAPHIVM
In Obitum
GERARDI MERCATORIS
AVI SVI, PIE AC PLACIDE VITA DEFVNCTI.

Amiſſum luxi qui nuper utrumque parentem,
   Heu miſer ad tumulos eogor abire novos.
Sic Ave, te venerande pium qui dogma dediſti
   Sæpius, hinc nobis te rapit atra dies?
Sic doctrina gravis, pietas ſic priſca peribit,
   Hæcque ſepulchralis conteget urna ſimul?
Abſit ut hæc periiſſe queant quæ Morte tenaci
   Conſumptos vitæ perpetis eſſe jubent.

Has liceat cineri lachrymas tamen addere Avito,
   Nam preſſus nimium, cor ferit ille dolor.
Thesbitæ ut quondam vatis ſucceſſor, in altum
   Quadrigis rapti dona recepit heri;
O utinam doctæ mihi ſic quoque portio mentis
   Cedat, & eximiæ pars quotacunque manus.

IOANNES MERCATOR,
ex primo-genito nepos, hiſce
mœrens parentabat.

The death of Gerardus Mercator at the age of 82 was recorded in the atlas, along with this laudatory poem by his grandson Johannes, who prayed that "a fraction of his learning might be bestowed upon me, or part of his distinguished craft, however small."

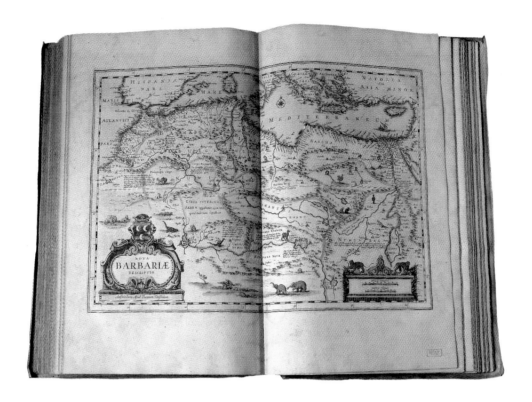

The Barbary States—Tripolitania, Tunisia, Algeria and Morocco—were in 1656 part of the Turkish Empire

JAN JANSSON published this edition of his *Nouvel Atlas*, or *Théâtre du Monde*, at Amsterdam in 1656. It was planned to consist of five volumes, four dealing with individual countries and a maritime atlas; ultimately it was extended to a six-volume work by the addition of an historical atlas. The Jansson atlas is in large part a continuation of the Mercator-Hondius collaboration, which first appeared in Amsterdam in 1606.

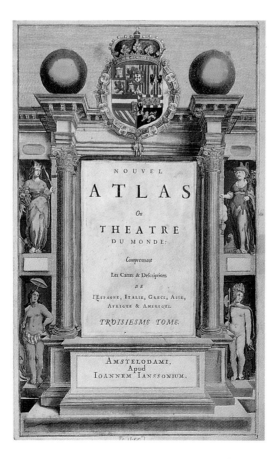

This atlas increasingly became the work of Jodocus Hondius and his son Henricus and then of Jansson, who married Jodocus's daughter. On the death of Henricus in 1650, Jansson acquired the stock and added the maps to his *Théâtre du Monde*, which had appeared in three volumes in 1639. A fourth volume of British maps, based on the work of Speed and Camden, was added in 1647; the final edition (1658–61) contained 11 volumes.

The elaborate arms of the King of Spain surmount the title page to Volume Three. This copy belonged to the Tyrolian Jacob Joseph, Count Wolchenstein, who signed and dated it in 1699.

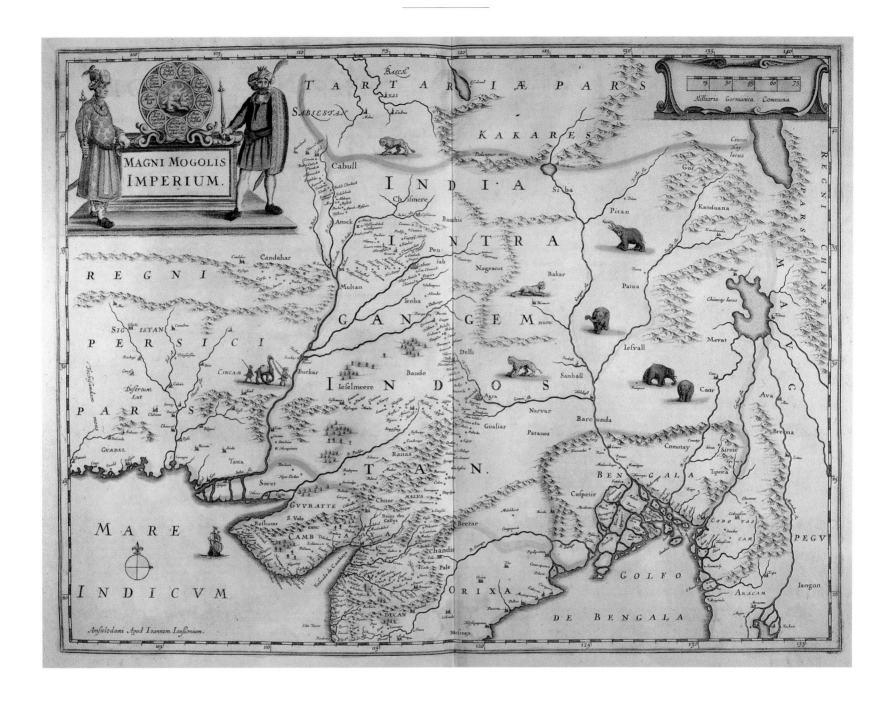

Ethiopia, or Abyssinia, was a name often applied to a vast area of the African interior. This fanciful cartouche suggests that it is part of the legendary Prester John's domains, a Christian empire believed to exist in Asia and Africa.

Europeans were fascinated by the power and wealth of the Mughal Empire, founded in the late 15th century by Babur, a descendant of the Turkish Tamerlane, in northern India and present-day Pakistan, and maps proliferated. Babur (r 1526–30) established a dynasty which effectively ruled much of northern India for two centuries.

The empire reached its apogee during the reigns of Akbar (1556–1605), Babur's grandson, who extended it from Afghanistan to the Bay of Bengal and south to the Deccan, and his son Jahangir. Jahangir's son, Shah Jehan, built many architectural marvels, including the Taj Mahal.

Jansson's map shows far more than the area of the Mughal Empire, even at its greatest extent, and there is some confusion with the empire of Genghis Khan which, some 200 years earlier, had flourished farther to the north.

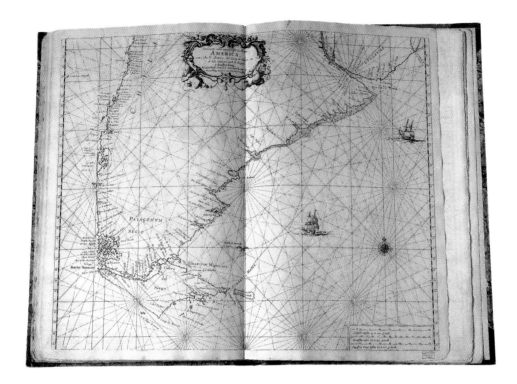

The importance of the map of South America lies in the accurate depiction of Tierra del Fuego as an island

B Y THE MID-1600S there were several sea atlases, and Hendrick Doncker's *De Zee ofte water-wereld* of 1663 contains important charts. It is introduced by a world map, *right*, dated 1660, compiled by Frederick de Wit who later used it in his own atlas. The map includes references to Australia, the outline of which is becoming recognizable, and to Abel Tasman's discoveries, but the island named for him appears as a single coastline, too far to the west. The mythical Terra Australis, which Mercator had inherited from Ptolemy, has vanished and California is drawn as a large island off North America.

The man on the right holds a cross-staff of Arab origin, with transoms which could be moved up and down. Eye-holes at the ends of the transoms allowed the top sight to be aligned with the sun during the day and the moon at night and the bottom sight with the horizon. The transoms were then adjusted until the sights lined up, when the angles of transoms and staff could be computed.

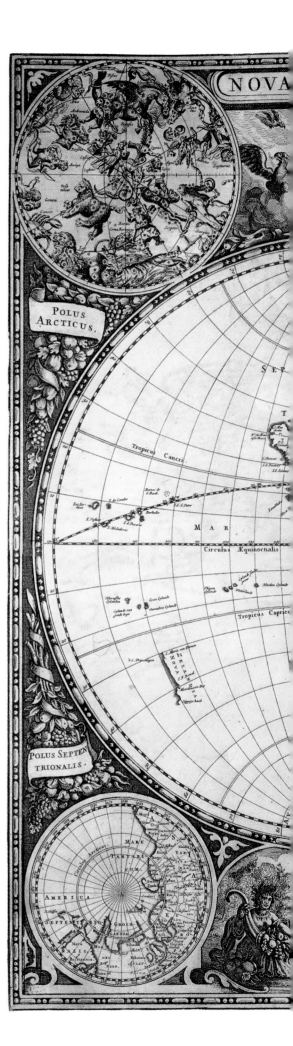

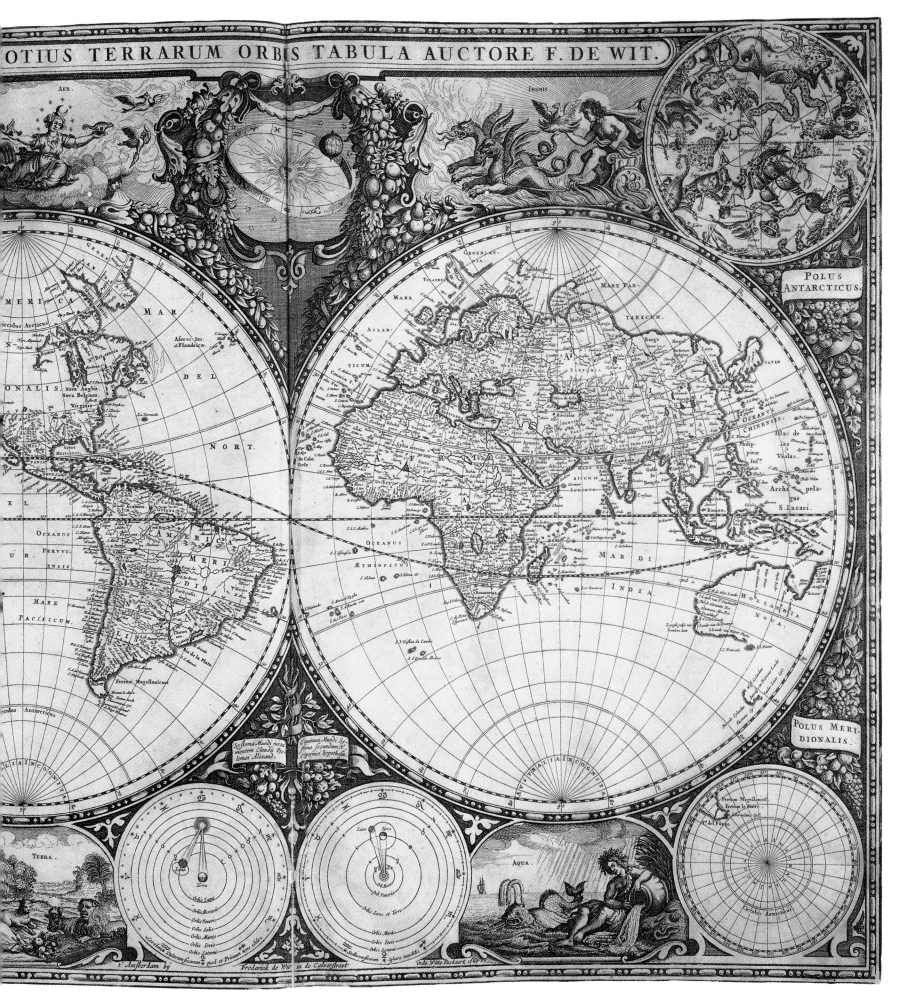

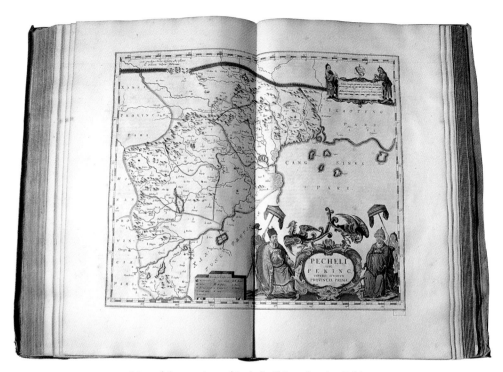

Map of the province of Pecheli, China, showing Peking

JOAN BLAEU was born in Alkmaar in Holland in 1598 or '99. The son of Willem Janszoon Blaeu, a globe and instrument maker, Joan went into partnership with his father and, when his father died in 1638, succeeded him as chart maker to the Dutch East India Company. He began his great publishing project, expanding the *Theatrum* from two volumes in 1635 to six by 1655, and his printing business became very successful.

Blaeu was invited to participate in public life, and he began to transfer the business to his son Pieter. In 1662 he held a vast sale of earlier stock, prior to the launch of the *Atlas Maior*. He was working on a further expansion of the atlas when his press in Gravenstraat, Amsterdam, was destroyed by fire and he lost his position on the city council, factors thought to have contributed to his death in 1673.

Borders on the American map show native peoples. The pair on the left are Peruvians; on the right, the king of Novae Albionis wears a feather crown.

The map of the Gulf of St Lawrence and Newfoundland, with its fishermen and lines of hooked fish, reflects the area's maritime richness.

In his map of India and part of southeast Asia, Blaeu shows the subcontinent too narrow, but Sri Lanka is beginning to assume its proper size and position. The Straits of Malacca are well surveyed because of their importance to ships sailing to China, and the southern part of Japan appears in the top right-hand corner of the map. The Korean peninsula is drawn as an island, and only a tentative scribble marks Australia.

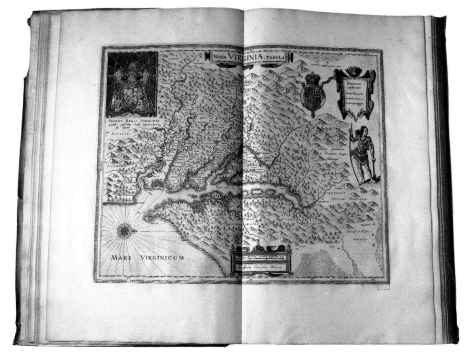

The map of Virginia, with the British royal arms, is based on that of William Smith. It is orientated with north on the right and marks Chesapeake Bay and the lands of the Indian tribes. An Indian, armed with a bow and club and carrying a dead animal, possibly a coyote, is shown and, at the top left, King Powhattan is depicted with a feather crown and smoking a pipe.

Pernambuco in northeast Brazil was first settled by the Portuguese, then occupied by the Dutch (1630–54). The map includes an illustration of the cultivation of sugar cane and the process of sugar production, a trade that attracted more Portuguese settlers to Brazil than to the rest of her extensive colonies combined.

The title page of Volume Twelve, *right*, is a fine example of typography from Blaeu's press; the decorative title page, *far right*, is that of Volume Eleven, the *Atlas Sinensis*, compiled by Martinus Martini. A Jesuit missionary in China, he prepared the 17 maps in this volume from Chinese sources. Together they form one of the most remarkable achievements of printing in the 17th century.

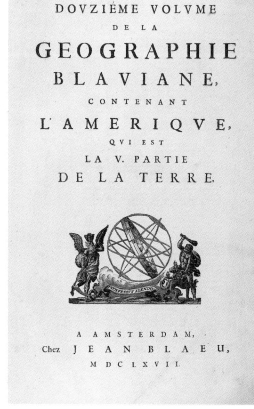

The "Celebrated Islands of the Moluccas", *right*, lie between the Celebes (Sulawesi) and New Guinea and yield a rich variety of fruits, and spices such as nutmeg and cloves. They were explored by Magellan in 1511–12 and occupied by the Portuguese; the Dutch took them in 1605. The map shows the wind roses and rhumb lines vital to navigators in such waters.

Blaeu's *Grand Atlas* was published in several European languages, as well as Latin. His printing house was the largest in Europe, with nine presses for printing text and six for printing plates. At its most active, in about 1660, the press employed some 80 men, and almost a million impressions were printed from around 1,000 copper plates of maps and town views. Blaeu was also the most important globe maker in Amsterdam.

The *Grand Atlas* was rivalled by that of Jansson; Blaeu kept the lead, adding a volume of 62 maps of Italy in 1640 and, later, a fourth with 59 maps of England and 49 of Scotland. The atlas is famed for its 600 splendid hand-coloured maps and large size, rather than its scholarship. Intended for the libraries of the aristocratic and wealthy, it was also deemed a suitable diplomatic gift. The fullest edition, the French *Grand Atlas*, appeared in 1663, the Spanish version in 1673.

A garland of tropical fruits, decorative and exotic, adorns the head of one of the maps of South America.

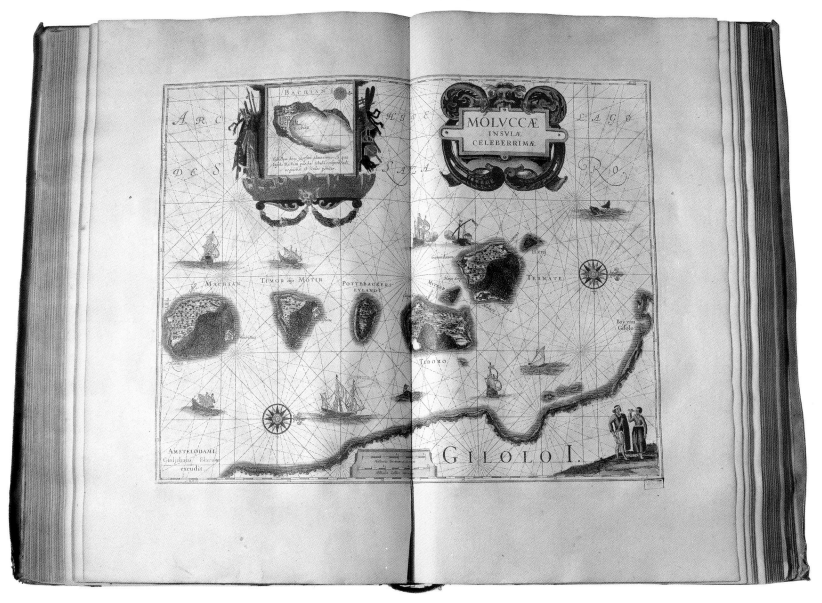

FREDERICK DE WIT (1630–1706) was a Dutch mathematician and cartographer who had been apprenticed to Willem Blaeu. In 1648 he founded a printing house in Amsterdam which produced land maps, sea charts and atlases noted for their fine workmanship and precision; his world map of 1660 first appeared in Doncker's atlas. He is best known for his *Zeekarten*, one of the most accurate sea atlases produced up to that time.

The title page of de Wit's general atlas of 1671 is baroque in style, but not very informative. The Titan, Atlas, who is usually depicted supporting the globe, here, correctly according to the legend, stands upon it and holds up the sky. He is shown straddling the "black rock" that Mercator believed marked magnetic north. The globe itself includes a fairly accurate outline of Europe, Africa, much of Asia and the western coastline of Australia.

Title page to de Wit's atlas of 1671

The Danish kingdom still included part of southern Sweden and Norway when this map was made. It gives a great amount of detail, including the sandbanks and shallows around the coast. Copenhagen, the capital, can be identified on the east of Zealand Island; it had not yet been extended to part of the island of Amager.

The views of Catania and Messina, *right*, are insets to de Wit's map of Sicily. Catania as seen here ceased to exist after the eruption of Mount Etna in 1669 and an earthquake in 1693. It was from Messina's sickle-shaped harbour that Don John of Austria sailed with his Christian fleet to defeat the Turks at Lepanto.

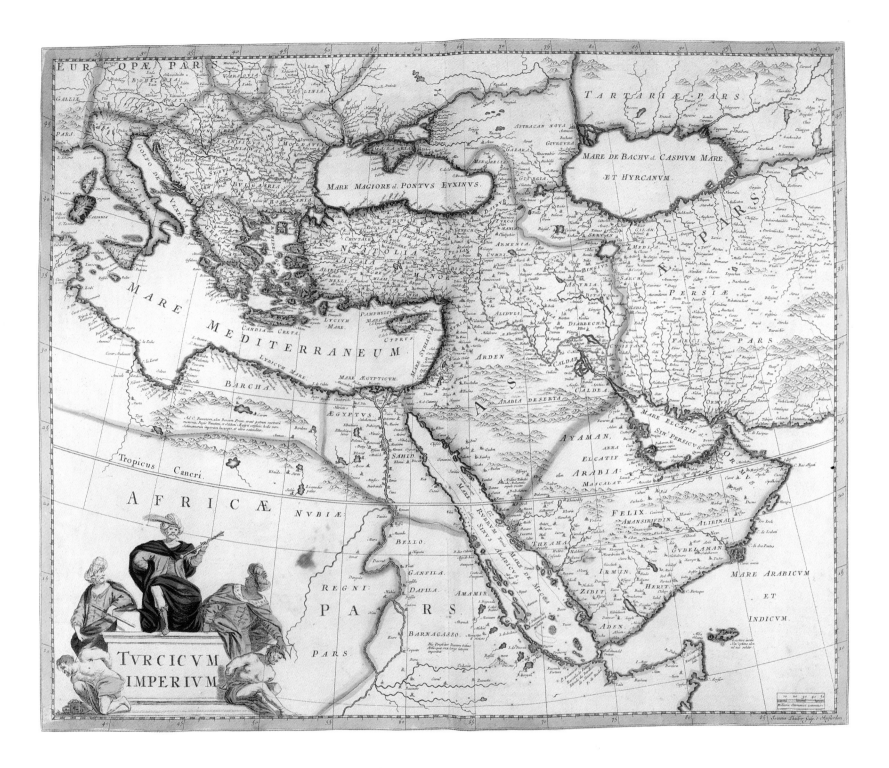

Despite the Holy League's naval victory over the Turks at Lepanto (1571), when some 10,000 Christian galley slaves were freed, the sultan's military might remained formidable. The European view of Turkish power in the late 1600s is reflected in the cartouche on the map of the Turkish Empire, *above*, which shows three Turks with two Europeans in bondage. The halt to Turkish encroachment in Europe, at Vienna in 1683, had yet to be called.

Although ill-defined, the empire's boundaries include the heartlands of Asia Minor, modern Syria, Iraq, Israel, Egypt, much of the Maghreb, Cyprus, Crete, the Peloponnese and Bulgaria.

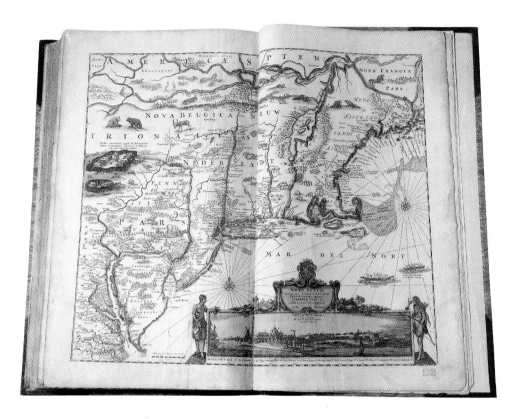

Map of Nova Belgica, comprising the northern parts of America

DURING THE seventeenth century, Dutch printers and engravers developed unrivalled skills. Among the most gifted of them was Justus Danckerts, whose maps were bought as much for their decorative qualities as for the information they provided. His atlas of *c*1710 was a typically lavish production.

The map of Nova Belgica (the north-eastern states of America), *above*, shows how European powers were attempting to allocate land. It includes a view of an Indian settlement, with stockade and grass-covered huts. The vignette of New York, earlier named New Amsterdam, has windmills and a Dutch appearance.

In this title cartouche, ornamented with figures of Arabs from the north and blacks from the south, Africa is symbolized by two sleepy-looking lions. Elephant tusks in the foreground are evidence of the growing importance of the ivory trade.

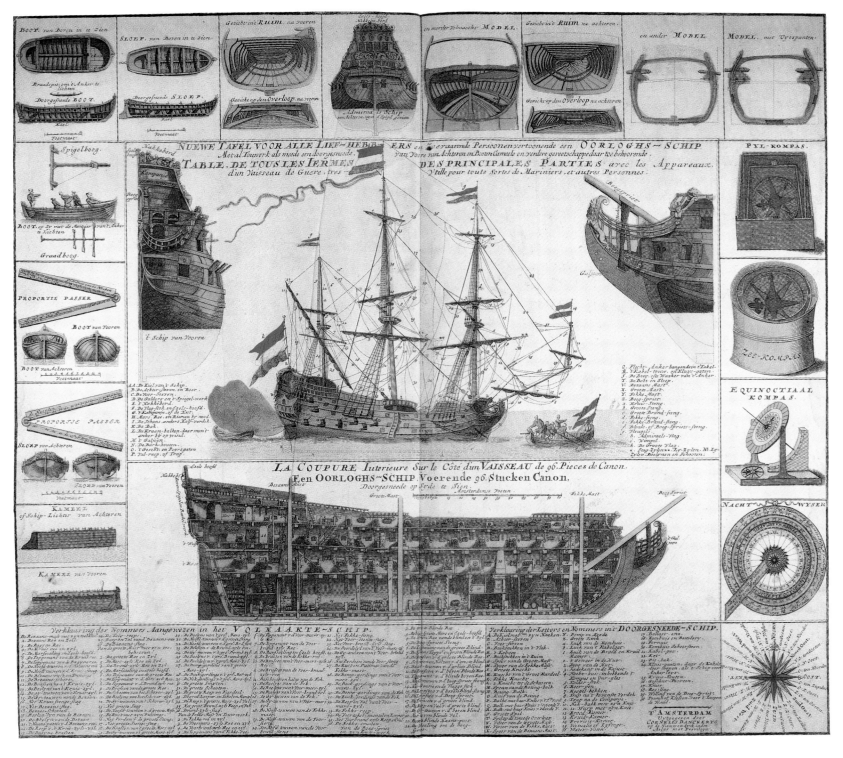

The cartouche for the "accurate and complete" map of the Turkish empire by Theodore Danckerts, the son of Justus, is embellished with typical figures, and the obelisk in the background is surmounted by the Turkish crescent. The lion from Africa and the black panther from Asia give an indication of the extent of the empire.

Detailed information about the construction of Dutch sailing ships is given on this plate, and the main feature is surrounded by illustrations of navigational instruments that were in use in the 17th century. An index at the foot of the plate identifies all the instruments as well as providing information about the parts of the ship.

Much of the explanatory text is in both Dutch and French, since these publications were aimed at an international market and the Dutch language was little understood outside Holland and its colonies.

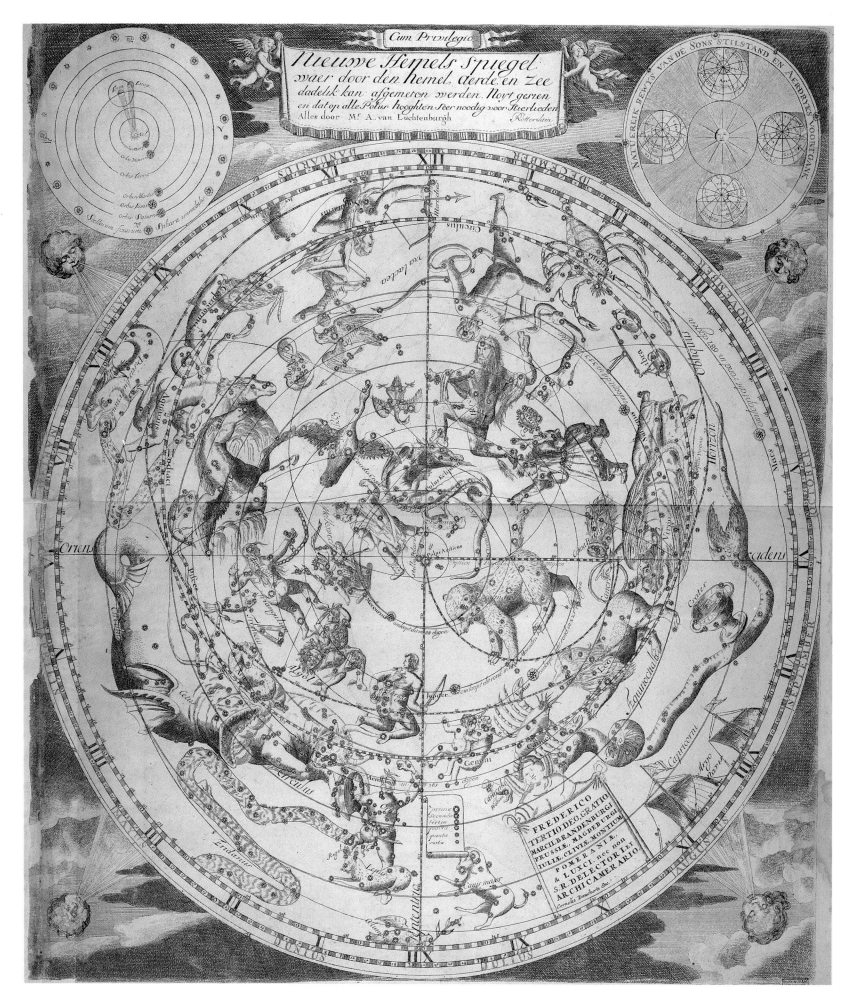

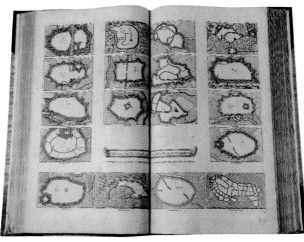

This highly decorative star map is bisected by the Equator, so the constellations in the upper half are those seen in the northern hemisphere, and those shown in the bottom half are seen in the southern. Lines on the map indicate how it could be cut into gores and applied to a globe.

Danckerts dedicated this map of the heavens, *left*, to Frederick III, Elector of Brandenburg. Only the constellations in the northern sky are given, but doubtless Danckerts intended to produce another map covering the southern sky. The inset on the left charts the orbits of the planets around the sun and of the moon around the earth; that on the right shows the progress of the earth around the sun.

Each of these 18 plans of fortresses in Lombardy is identified by name; many were not truly fortresses but cities occupying strategic sites, such as Venice and Genoa. They were published by Cornelis Danckerts, the son of Justus.

PIETER SCHENCK 1706

PIETER SCHENCK (1645–1715) operated a printing and engraving business in Amsterdam. He was not among the great cartographers but published several atlases, including the *Atlas Contractus* in 1705, the *Théâtre de Mars* a year later, and the *Atlas Minor*, illustrated here, *c*1715.

After his death, his son, also called Pieter, continued publishing maps, much in his father's style. Later he went into partnership with Gerard Valk who, like Pieter the Elder, had acquired plates from the workshops of Blaeu and Jansson, but Dutch map makers had by now been overtaken by the French.

This copy of the *Atlas Minor* belonged to the 2nd Duke of Newcastle (1720–94), whose arms are embossed on the cover

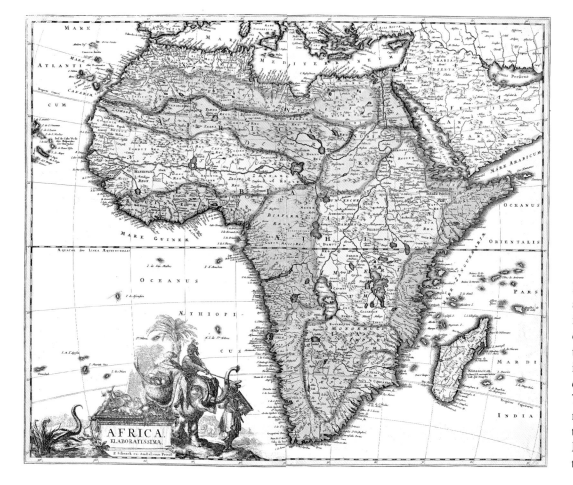

The shape of Africa is now accurately depicted, but Ptolemy's lakes of Zaire and Zaflan still appear, although Guillaume Delisle had not shown them on his African map of 1700. Some of the ancient tribal regions, such as "The Kingdom of Biafra", are marked.

Estuaries of major rivers appear, even if wrongly identified: the Senegal is carefully delineated, but the river runs due east and is called the Niger, whose course is quite inaccurate. The Nile, which Ptolemy rightly believed rose in the Mountains of the Moon, is shown too far to the south.

A bird's-eye view of the sound looking toward Copenhagen on the island of Zealand adorns the title to the map of the Danish Islands. On the right is Elsinore, with the fortress-palace of Kronburg, the setting for Shakespeare's *Hamlet*; on the left of the sound is the city of Elsingburg.

The map of western Europe has been used to show how a total eclipse of the sun occurs and how it is seen as such in only a small part of the globe. Lines to the left and right show those areas where it is seen as a partial eclipse. An inset between Iceland and Norway gives details of the total eclipse that occurred on 3 May 1715. Although the phenomenon was understood even in Greek and Roman times, it is interesting to see it graphically demonstrated in an atlas of this period.

A map of the Mediterranean includes this vignette of two men bearing a heavy harvest of grapes. Below, various scales are shown for measuring distances on the map, including *Millaria unius horae*: the distance a Roman soldier could march in an hour.

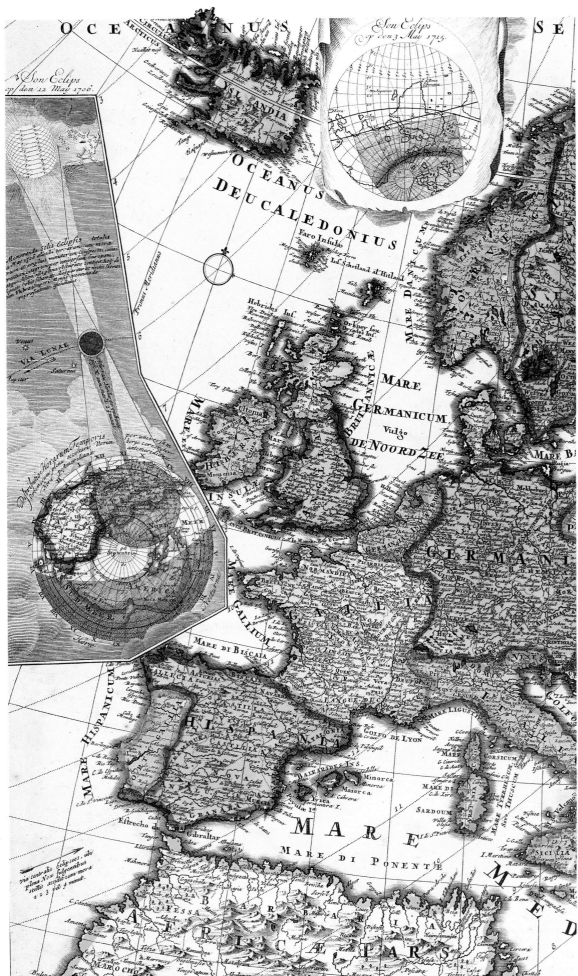

Moses is pictured receiving the tablets of the Ten Commandments in this dramatic title cartouche to a map of the Holy Land which shows the regions occupied by the twelve tribes of Israel. The map was compiled by Johann Baptista Homann of Nuremberg.

Schenck's world map, dated 1706, is presented in two spheres on the Mercator projection. The western hemisphere shows the Americas (with the north coast still unexplored), the east coast of the Gulf of Carpentaria, part of Van Diemen's Land (Tasmania) and some of the west coast of Zealandia Nova (New Zealand). The sphere covering the eastern hemisphere includes all of Africa and Asia and the greater part of Australia, here called Hollandia Nova. The two spheres are surrounded by eight smaller ones depicting different parts of the world.

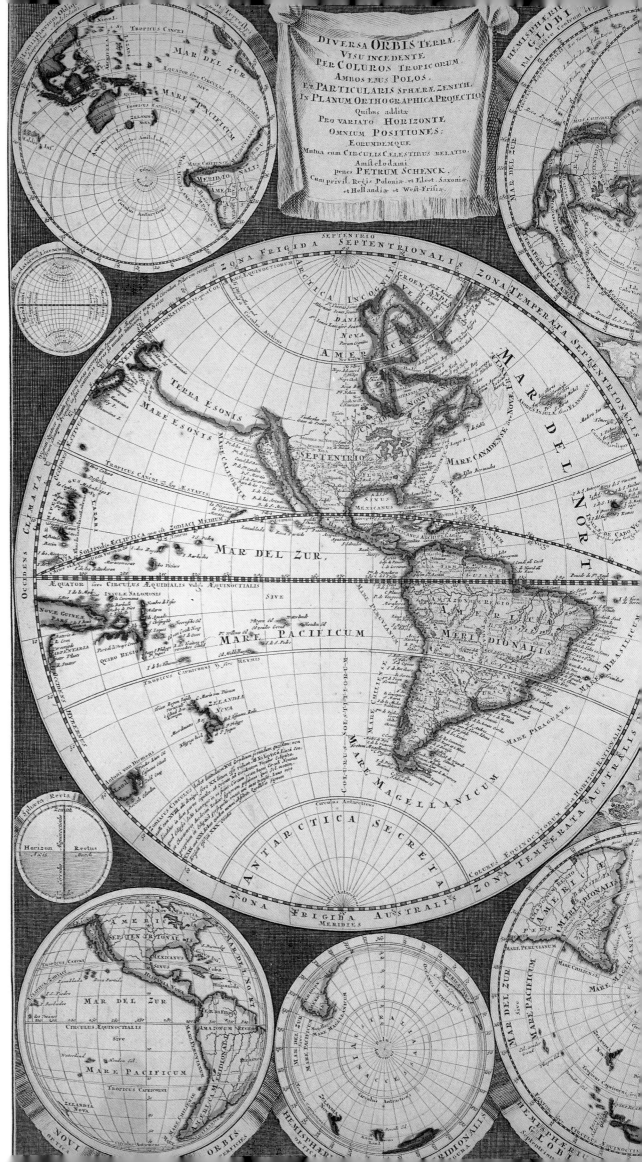

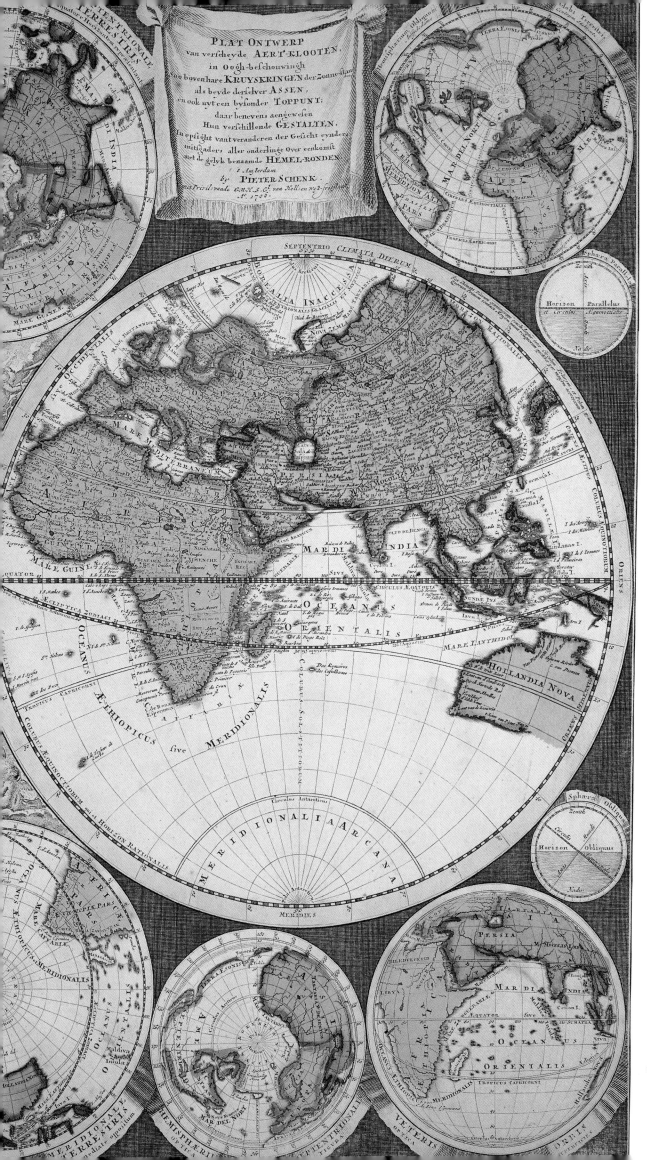

The great Blaeu publishing house was ruined in 1673 when a fire consumed the press and most of the stock. Blaeu died in the following year, and such copper plates as survived were sold to Pieter Schenck. He further increased his resources by acquiring printing plates from the descendants of Claes Janszoon Visscher, whose map-publishing company had expanded over three generations.

Schenck was now well equipped to publish a handsome atlas, even if its information was often out of date. This was not unusual, since there had always been a wide divide between the few highly important, influential works and their numerous cheaper imitations, many of which were produced for display, rather than for practical use.

The Holy Land map is further embellished with an illustration showing bearded patriarchs pointing to an unrolled map of the Red Sea and Egypt. This vignette, an allusion to Moses leading the Israelites through the Red Sea, balances that of Moses the Lawgiver.

CHRISTOPHER SAXTON 1579

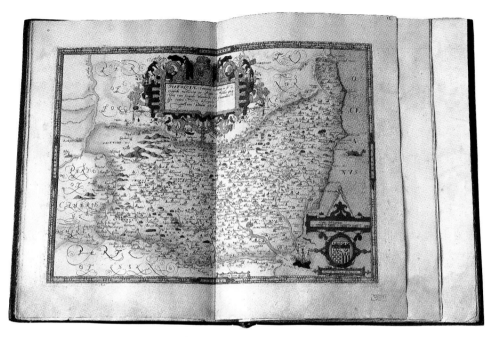

Map of Suffolk, showing the county arms

THE FIRST MAPS accurately identifying the towns and villages of England were produced by Christopher Saxton between 1574 and 1579. They were originally available as 34 county surveys, but the final atlas included a general map of England and Wales. During the time Saxton spent surveying, drawing and preparing the maps for publication, he was helped financially and with introductions to the local aristocracy by his patron, Thomas Seckford, Master of Requests at Elizabeth I's Court.

All the plates in this copy of the atlas were hand coloured and it would have cost about twice as much as the plain version. The copper plates from which the images were printed were engraved by the most gifted of the English and Flemish craftsmen. After the publication of Saxton's atlas, all English maps of the period were based on his; he is, without doubt, the father of English cartography. In the county map of Suffolk, *above*, the important port of Harwich is given due prominence.

A fine portrait of Queen Elizabeth I, brilliantly illuminated with deep colour and gold, forms the frontispiece. The inscription is a tribute that claims "while much of the world flounders in blind delusion, you bless the British people with enduring peace, wisely and justly holding the reins of government".

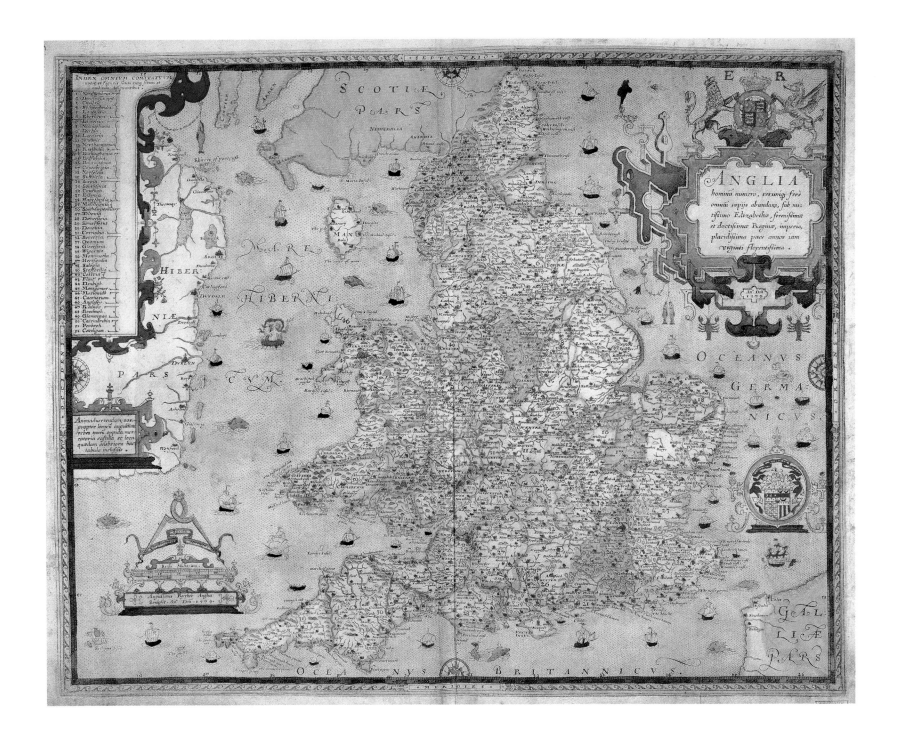

The coats of arms of the 21 English counties, together with the personal arms of some of the aristocratic families who helped Saxton, appear on one page of the atlas. Opposite is a list of the counties, detailing the cities, castles and other important features.

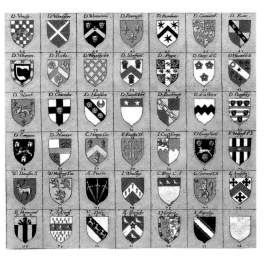

The general map, which contains the essentials of the individual county maps, gives a fascinating picture of the realm before the union with Scotland in 1603.

A key lists 52 counties and cities, many of them with Latinized names: Cornwall appears as Cornubia, Hampshire as Southamptonia. The arms are those of Thomas Seckford, Saxton's patron.

This cartouche, which appears on the map of Wiltshire, here given its Latinized name of Wiltonia, is surmounted by the royal arms of Queen Elizabeth; those of Thomas Seckford appear below.

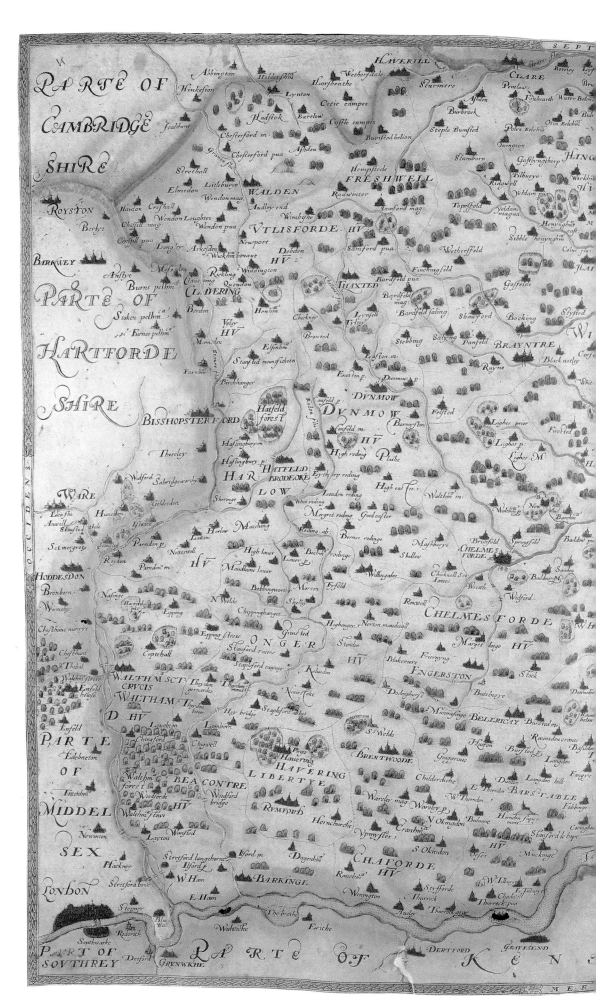

96

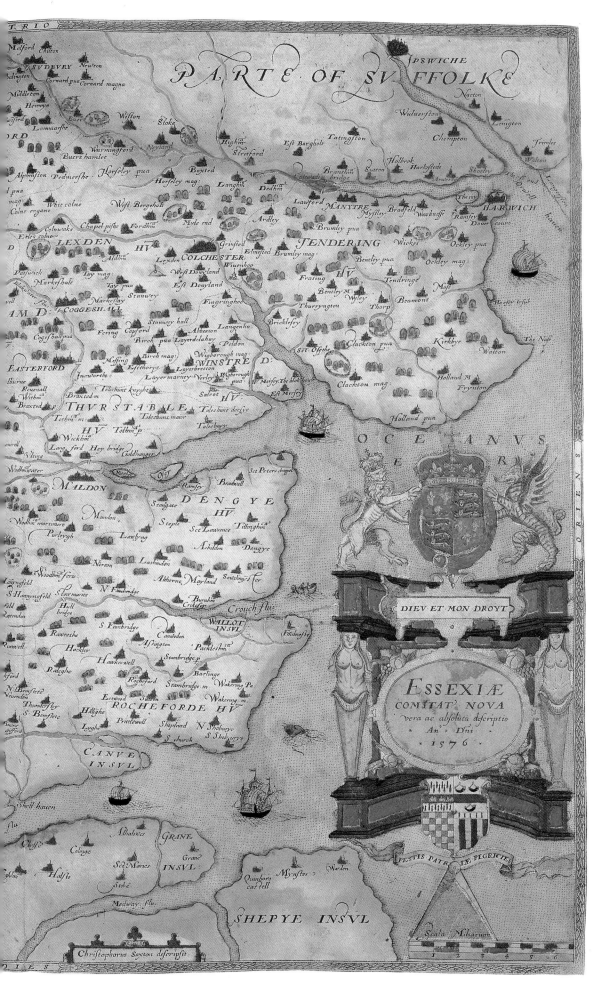

The title cartouche for the plate of Warwickshire and Leicestershire, once again with the royal arms, states that the cities, villages, rivers and other important features are faithfully recorded.

Saxton's map of the county of Essex, shows London in the left-hand corner of the map, with the old London Bridge connecting Southwark with the city. Hatfield House, where Queen Elizabeth spent some of her early years in virtual captivity, is clearly marked, as are the many navigable rivers of the east coast. These were important to England's trade with Europe and were busy with sailing ships carrying wool from England and bringing back finished cloth from the Low Countries and wine from France.

JOHN NORDEN (1548–c1625) was born in Somerset and trained as an attorney; but around 1580 he started working as a cartographer, surveying large estates. In 1588 he undertook a survey of Northamptonshire and completed a map of the county three years later. Despite seeking the patronage of Lord Burleigh, Queen Elizabeth I's powerful Secretary of State, he was unable to raise sufficient funds for his *Speculum Britanniae*, and maps of only five counties were printed.

The title page, *left*, to the first part of the *Speculum Britanniae* (1593) covers the county of Middlesex. The guide provides an historical description of the county, giving details of its cities, towns and parishes. Norden introduced a number of innovations into map making, including a triangular table for giving distances between towns, and his were the first printed English maps to show roads and to indicate a scale of distances. Several of these maps were later used by William Camden in his *Britannia*.

The elaborate, coloured map of "Hartfordshire" and fully coloured and gilded arms of Queen Elizabeth in this copy of the *Speculum* (1598) suggest that it was meant for presentation to the queen. Hatfield, where William Cecil, Lord Burleigh, was shortly to build his great mansion, is in the centre of the map.

Norden's interest in the county's history is evident from the note, south of Luton, which reads "a battle where the Danes were overthrown"—at a place called Dane End.

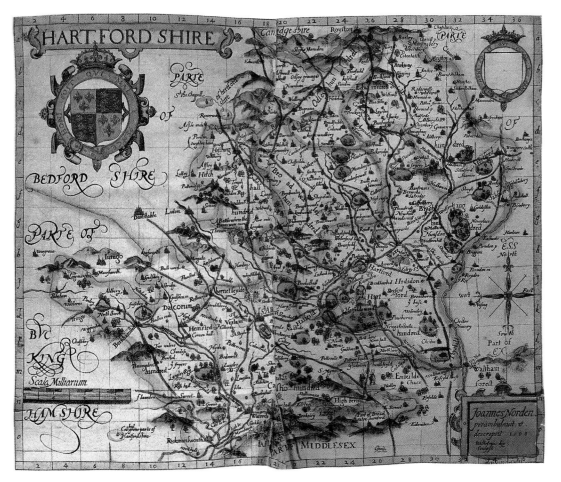

Queen Elizabeth's arms, which appear in the Middlesex volume of the *Speculum*, include a quartering with the letters SPQR—*Senatus Populusque Romanus*—an oddly republican sentiment to associate with so absolute a monarch.

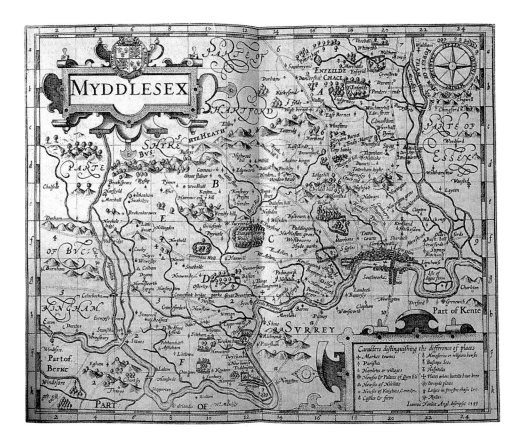

The aerial view of the City of Westminster, *below*, appears in the Middlesex volume. At bottom left, south of the Thames, is Lambeth House; on the opposite bank is the old Palace of Westminster, now the site of the Houses of Parliament. Westminster Hall is identified, but most other buildings no longer exist.

St James's Park, with many deer, Charing Cross and Covent Garden are shown. Along the river, south of the Strand, are the great town houses of some of the nobility.

Norden's county maps are distinguished by special symbols to indicate the size and importance of particular places, as well as individual buildings such as castles and the houses of the nobility, hospitals and mills. In the map of Middlesex, *left*, the symbols are, for the first time on any map, contained in a key.

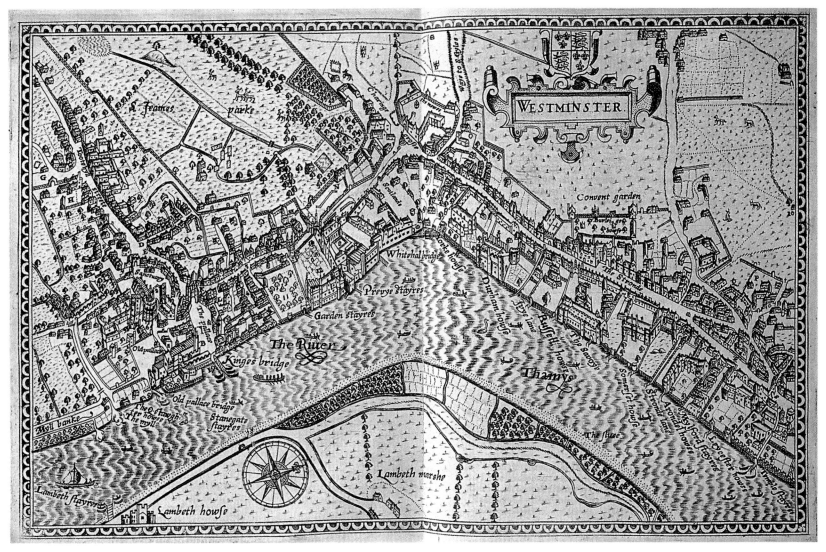

Dedication to King James I in the 1614 edition of Speed's atlas

JOHN SPEED (1542–1629) was born in Cheshire but spent most of his life in London, working as a tailor. His knowledge of antiquities drew him into the circle of Sir Fulke Greville and Sir Henry Spelman, an antiquary, who helped him publish 54 maps of England and Wales in 1608–10; these were incorporated in the *Theatre of the Empire of Great Britain* in 1611. Speed contributed to the Mercator-Hondius atlas, while Hondius engraved the plate, *above*, showing King James I's arms for the 1614 edition of Speed's maps. A manuscript note records that a previous owner paid £1.11.6d for this copy of the atlas, a considerable sum in 1628.

Portraits of early kings (Hengist, Offa, Edwin and Ethelbert among them), views of the kingdom each ruled, and a grand title cartouche decorate a map of Britain "in the tyme of the English Saxons especially during their Heptarchy".

The map itself shows the supposed seven kingdoms of the Angles and Saxons in the 7th and 8th centuries. Historians anxious to establish an illustrious ancestry for the English monarchy firmly believed in this near-mythical past.

The map of Montgomeryshire was based on Saxton's; the arms are those of Philip Herbert, Earl of Montgomery, who probably paid for the plate to be made. Inset is an aerial view of the walled county town and its castle; the public well is at the top of the High Street and, to its right, the stocks. The layout of the town centre is little changed today.

Like others on the Welsh border, the county had many castles, built in medieval times to protect the English against raids by the Welsh.

In the text preceding each county map Speed described it and summarized its history. On the back, he gave an alphabetical list of the Hundreds (sub-divisions of the county) and of the most important features marked on the map—towns, churches, castles, rivers, hills.

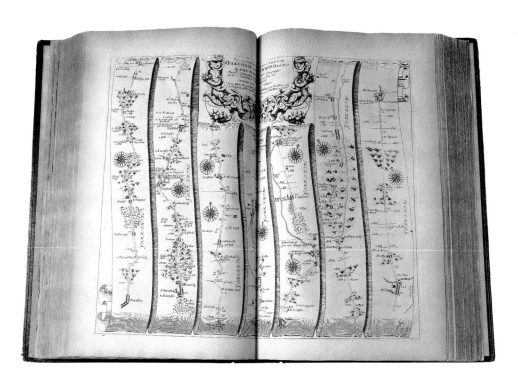

The route from Oakeham in Rutland to Richmond in Yorkshire

Mermaids support the royal coat of arms and hold banners bearing maps of London and York, England's main cities, above the introductory text, and aquatic putti romp in the sea.

The capital "G", *right*, encloses a globe depicting Europe, Africa and Asia.

THE FIRST ROAD ATLAS was produced by John Ogilby, a Scottish writer and printer, who in 1666 was appointed Royal Cosmographer, with the task of surveying all the roads of the kingdom. The resulting volume, *Britannia*, was engraved in strips on 102 copper plates and printed by Ogilby himself in 1675.

Though decorative, the maps were meant primarily for use by travellers. The section, *above*, covering a distance of 87mls/140km, with the compass changing orientation on every strip, is typical. A small pocket version of the library atlas was also produced, but few examples survive since they fell apart with constant use.

Leaving London, which is shown in detail, the road to Banbury, *below*, leads first through nearby villages, such as Acton, in Middlesex, then into the counties of Buckinghamshire and Oxfordshire, whose boundaries are marked by a dotted line. The map includes relevant crossroads, villages, churches and the mansions of the rich. Because of the poor state of even main roads, small brooks are marked for, swollen after heavy rain, they could become a hazard.

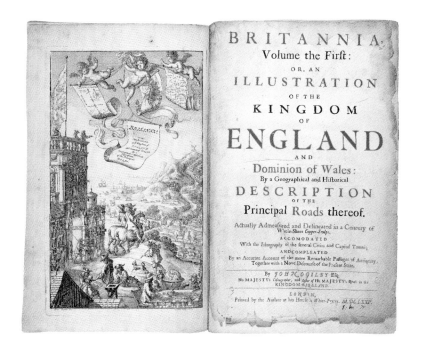

The frontispiece to the atlas was designed by Wenceslaus Holler (1607-77), the distinguished Bohemian engraver whom Charles II appointed "His Majesty's Designer". In the foreground, a group of men studies a globe, while others examine one of Ogilby's maps. Behind them two horsemen, equipped with their route maps, set out from London—the city's coat of arms appears on the flag above the gate—into an idealized countryside.

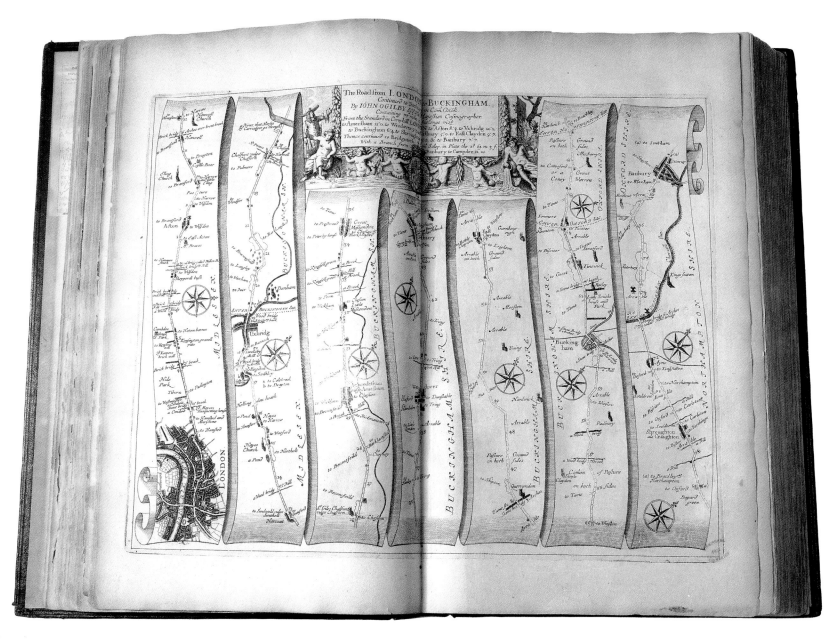

ATLAS

MARITIME

Non pius Æneas, non sic evasisset Ulysses
Carpsisset rectum, per mare Lybys iter
Si duxisset Atlas, latraret Scylla, Sirenes
Cantarent: frustra, carbasa tuta forent.
Per vada, per syrtes, caecosq; ad littora rupes
Edicet Oceano quoslibet ire vias.
Dißde Vicino, Gallus Tritonia regna,
Quæ Batavum, captat; Marti vel arte magis.
Hic atarum reparato decus, nec cedite Galle
Floreat Herculeus Dædaleusq; labor.

Romanus de Hooghe J.U.D. et
Com Reg. tab hanc sui
Utrecht sanci. et inv. 1693

A AMSTERDAM chez PIERRE MORTIER.

CHAPTER FOUR

The AGE of EMPIRES

THE FIRST EUROPEAN colonial possessions were established in the early sixteenth century, following the voyages of discovery by Portuguese and Spanish navigators. These embryo empires gave rise to the maps discussed here

The decisive decade of exploration was the 1490s and, by the second quarter of the sixteenth century, Portuguese voyages around the southern tip of Africa and into the Indian Ocean, and Spanish journeys across the Atlantic to the Americas and thence into the Pacific, led to the circumnavigation of the globe. The information gleaned during these stupendous undertakings made it possible to compile maps which, though inaccurate in important details, showed what the world was really like and thus set the stage for the imperial aspirations of the contending powers.

At this time, too, the attractions of the lucrative spice trade began to draw the English into the Indian Ocean and southeast Asia and thus into competition with the Dutch and Portuguese. This expansion of commercial interest and the foundation of the English East India Company in 1600 were intended to break the

Atlas, in the opening plate of Pené's atlas, opposite, is seen holding up the sky to prevent it falling; in the title page to Renard's atlas, below, in a common, but incorrect, representation, he is shown holding up the earth.

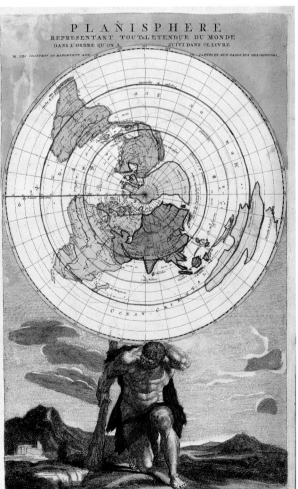

monopoly of the Dutch in the region, but once they were actually in India, the English soon discovered the profitable occupations of tax collecting and administration.

A base established at Amboina (Ambon) in the Moluccas was overrun by the Dutch in 1623, and by the end of the century the Company had decided to concentrate on the west coast of India, leaving the islands to the Dutch.

All trade in the new English colonies was, after 1651, tightly controlled by a series of legislative acts, for the colonies were now seen to a large extent as valuable economic appendages of the home country, to be exploited at will. Convicts were also sent to the new territories in increasing numbers, in an effort to boost their European populations.

The end of the threat of Stuart rule in 1689, and the accession of the staunchly Protestant Hanoverian monarchy, heralded a return to hostility to France, and for the next 120 years England was either at, or preparing for, war with France. Each of the seven conflicts between 1690 and 1815 affected the colonies, and by the end of that time France had lost all her overseas possessions.

Like England, France had not participated in the early voyages of exploration, but it was not slow to try to reap the benefits and, in common with England, looked to the North American continent and to India. Colonies were established in what is now the United States and Canada and trading stations were set up in India.

This fledgling empire was shattered by the War of Spanish Succession (1702-13), which allowed the English to assert themselves militarily in Europe and to expand their possessions by taking over the smaller colonies of France and other rivals by conquest.

After the Treaty of Utrecht (1713), by which Britain gained St Kitts, Newfoundland, and unchallenged title to the Hudson Bay area, Britain enjoyed peace for almost 30 years, and its merchants took advantage of the concessions that had been gained. The colonies in North America expanded, and slaves were transported and sold to Spanish America at a good profit and to plantations in the Caribbean and North and South America, where a constant supply of cheap labour was required for growing and processing the all-important sugarcane.

France still aspired to a leading role in world affairs, however, and a further conflict with Britain over the Austrian succession in 1740–48 soon extended their respective possessions in North America and India, where religious, commercial and racial factors also contributed to the bitter struggles.

At the end of the Seven Years' War (1756-63), the might of metropolitan France was broken, and it lost most of its

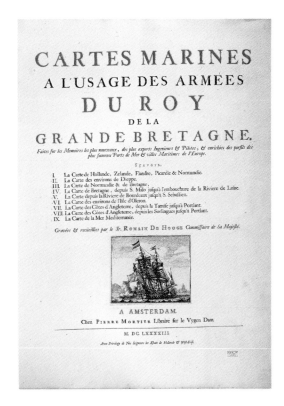

The title page, *top*, of Pené's marine atlas, which contains nine charts of the European coast; Renard's nautical atlas covers the whole world.

colonies. The trading stations in India were restored, but France's power was gone, and Britain emerged as the leading imperial nation. The removal of the French presence in North America contributed to the growing confidence of the American colonists, who soon challenged British authority and in 1782 won their independence.

The loss of its American colonies was a severe blow to Britain, but loyalist refugees built new settlements in Canada, and the Napoleonic Wars allowed Britain to add new territories such as Ceylon (Sri Lanka), Trinidad, and the Cape of Good Hope to their dominions. By this time the motives for acquisition were often strategic rather than commercial, and occasionally even more bizarre reasons prompted settlement; among these was the founding of the fully fledged penal colony in 1788 in Van Diemen's Land (Tasmania), the island off the southern coast of Australia.

The death of the Indian emperor Aurangzeb in 1707 not only signalled the beginning of the end to the Mughal Empire but allowed the British East India Company further to extend its power in the country. Conflict in India between the English and French was inevitable, and in the middle of the eighteenth century the Company ceased to rely on the power of the Moghul emperor to keep the peace.

A 25-year-old clerk, Robert Clive, was sent to confront the French in 1749, and by 1761 they had been effectively eliminated, as indeed had the real power of the Mughals. The British began to penetrate the interior, but this vast colony did not come under the British Crown until after the Indian Mutiny in 1857. The establishment of the outpost of

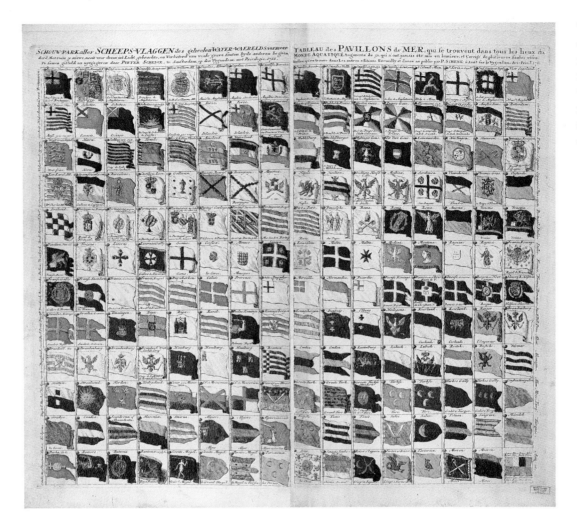

The cosmopolitan appeal of atlases at this time is indicated by the headings in Dutch and French; while Pieter Schenck's name points to close cooperation between the cartographers.

The splendidly colourful pages from the centre of Delisle's atlas depict national and city flags from all parts of the world, including the Papal States and Jerusalem.

Singapore in 1819 created a base which served the trading stations that had been established in southeast Asia and along the coast of China. The whole colonial edifice was administered and controlled from the Foreign Office in London.

The long years of comparative peace during the nineteenth century allowed the steady growth of Canada, Australia and South Africa, and a fair degree of self-government was soon introduced. This was followed by the creation of federal unions, the first being that of Canada in 1867, with Australia in 1901 and South Africa only in 1910.

The considerable influence exerted by the imperial theme, particularly since about 1800, is demonstrated in the way the atlases have been compiled. The glorification of a nation has often gone hand in hand with the accurate delineation of the topography of a particular area, but maps have also been used to show false boundaries, which could subsequently be disputed to a nation's advantage. The hard work, skill and dedication of the navigators, explorers and cartographers are rarely in doubt, but the objectives of those who commissioned and published the atlases is frequently suspect.

Contents page for Petty's *Hiberniae delineatio*

I N 1654 Sir William Petty initiated a survey of Ireland by instructing ex-soldiers in the science of taking bearings. The Cromwell government had decided to confiscate the property of all who could not prove their loyalty to the Commonwealth; the money gained would pay its many creditors, but it was necessary first to survey the estates involved.

After Charles II's restoration to the English throne (1660), Petty was encouraged to continue his work despite his Cromwellian connections. The results—surveys rather than topographical maps—covering the Irish parishes on a scale varying between 3–6in/7–15cm:1ml/1.6km, were used by Petty as a basis for his atlas, *Hiberniae delineatio*, in 1685.

Petty's atlas contained this folding "General Mapp of Ireland", as well as maps of individual counties arranged under the provinces of Leinster, Munster, Ulster and Connaught. It was the first to be based on a direct survey and became the source for most maps of the country for more than a century.

This vignette shows putti as surveyors—a far cry from ex-soldiers. One uses a pair of dividers, while the other records the dictated information; the scale in English and, interestingly, slightly longer Irish miles is set out below.

All the county maps are preceded by provincial maps, such as this of Ulster. Each map was given a compass rose and the topographical details include rivers, lakes and mountains. The city of Belfast, situated at the mouth of the Lagan River on Belfast Lough, an inlet of the North Sea, is marked, but Antrim was the more important place at the time.

CHARLES PENE 1693

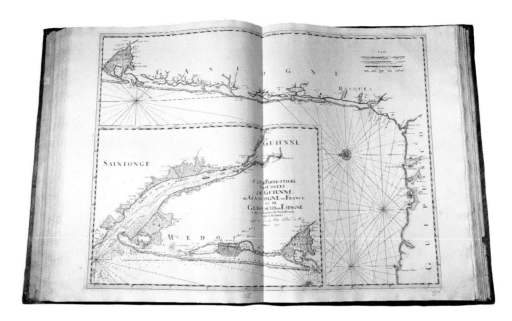

The Gascony coast with an inset of the River Gironde in *Le Neptune françois*

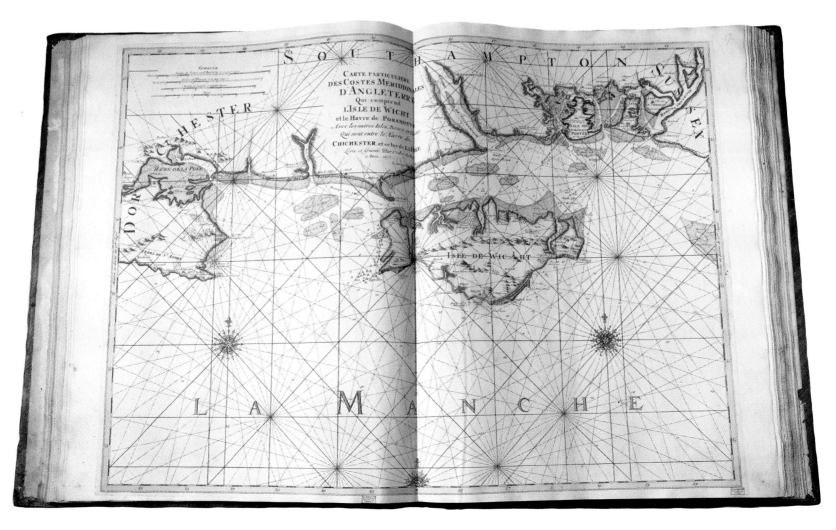

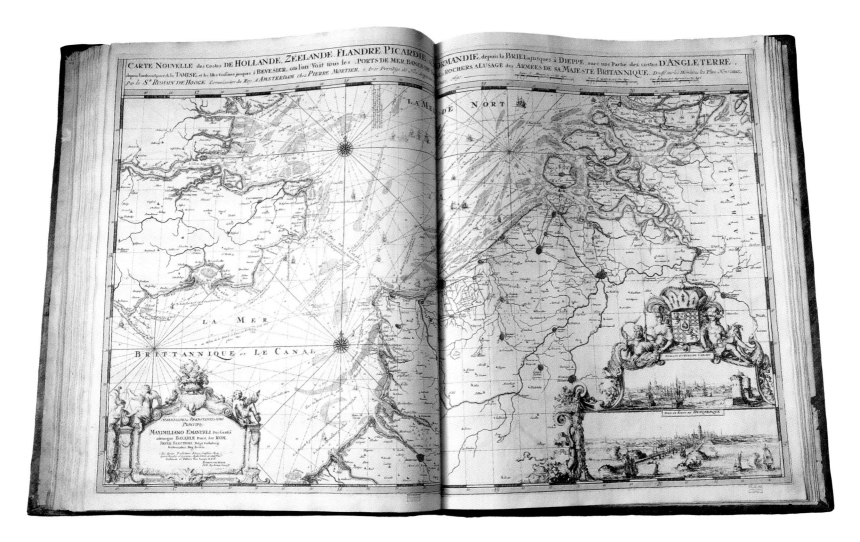

I N THE LATE 1600s, the charting of French coastal waters was undertaken by the Ingenieurs du Roi, and the results, edited by Charles Pené, were published in Paris in 1693 under the title *Le Neptune françois*. In the same year, another edition, with re-engraved maps, was issued in Amsterdam by Pierre Mortier. It contains nine charts of the French and Dutch coasts, two of the English south coast and a folding map of the whole Mediterranean. The two editions are bound together in the Birmingham collection.

This plate, in Pierre Mortier's *Cartes Marines*, covers the northern English Channel and the coasts either side of it. Interestingly, it states that it was compiled "for the use of the armies of the King of Great Britain".

The hydrographic map, *left*, of the south coast of England, includes the Isle of Wight, Chichester and Portsmouth. This was given great prominence because smaller ships were able to negotiate the shallow channel leading to it.

The inclusion of wind roses and rhumb lines is somewhat oldfashioned, but the map is extremely accurate, and shoals, sandbanks, soundings, beacons and anchorages are carefully indicated

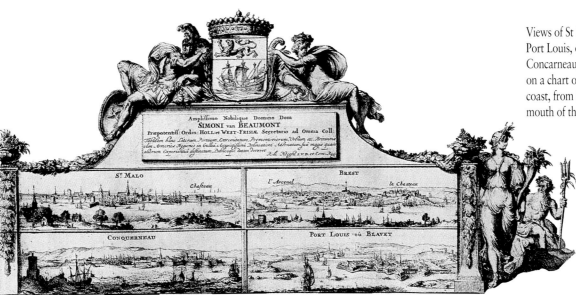

Views of St Malo, Brest, Port Louis, or Blavet, and Concarneau, *left*, appear on a chart of the Brittany coast, from St Malo to the mouth of the Loire.

111

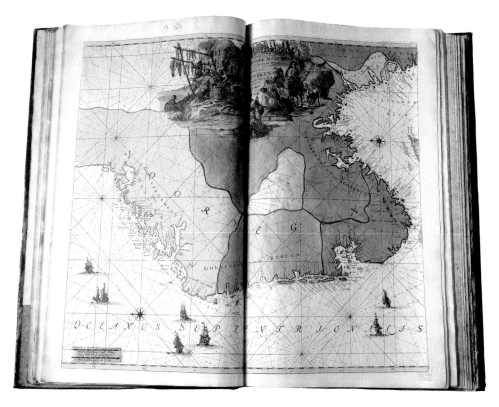

Mariners' map of Norway, with wind roses and rhumb lines

Orientated with north at the left, this map covers the coastal area of Brazil from the estuary of the Amazon in the bottom left-hand corner to the River Plate at bottom right. The interior is quite blank.

The Amazon remained unexplored by Europeans until the 19th century, but the Plate was navigated and partially settled by this time; Rio de Janeiro and San Salvador were already thriving cities.

The sumptuous title illustration shows the Spanish Conquistadors and a wealth of local fauna: alligators, armadillos and llamas among them.

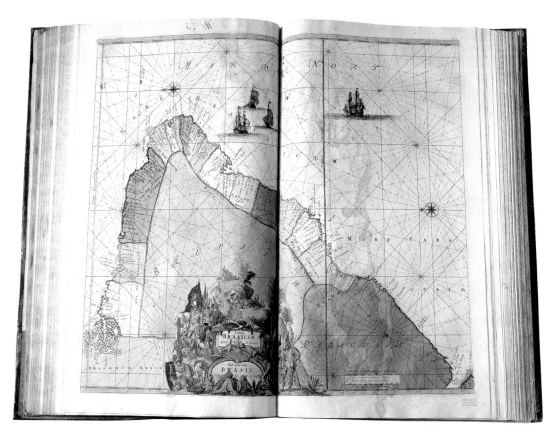

The decorative dropped capital "S" of the word "Sire" begins Renard's dedication of the atlas to King George I of England, a fine portrait of whom was specially engraved for the atlas.

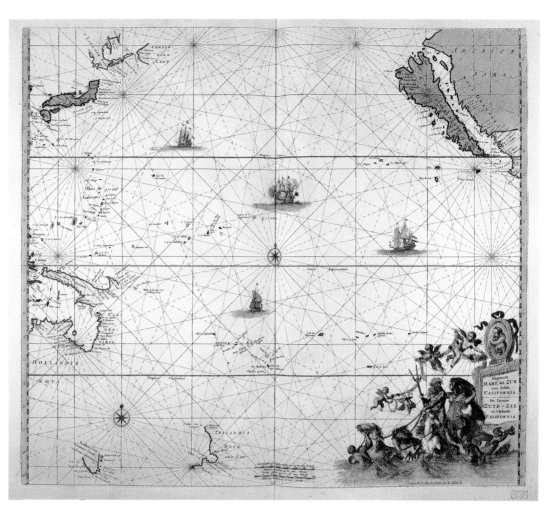

The map of Europe includes part of north Africa, where these two elephants appear. They probably represent the Atlas Mountain elephants used by Hannibal, which later became extinct.

The Pacific map shows Australia and New Guinea—incomplete— and, to the north, the main Japanese islands. It also perpetuates the discredited theory that California was an island.

RENARD SEEMS to have entered cartography by his marriage to the daughter of an engraver, Daniel de la Feuille, who had issued an *Atlas portatif* in 1701 and 1706. Renard's *Atlas de navigation et du commerce* was first published in Amsterdam in 1715, two years after the Treaty of Utrecht gave Holland its independence. In recognition of the important role played by Britain in the defeat of France and the restoration of liberty to the Netherlands, the atlas is dedicated to King George I.

The 28 hand-coloured maps, intended for use by navigators, have wind roses and rhumb lines. Renard's atlas was reissued in 1739 and 1745.

India and Ceylon appear at the bottom left, and Hollandia Nova (Australia) at the top right of the map of the East Indies, which is orientated south to east. Although the East Indies had been well mapped for more than 200 years, and by now the Gulf of Carpentaria had been thoroughly explored, the western half of New Guinea is much exaggerated and the Japanese island of Kyushu is still misaligned. The vast extent of the Pacific is, however, now recognized.

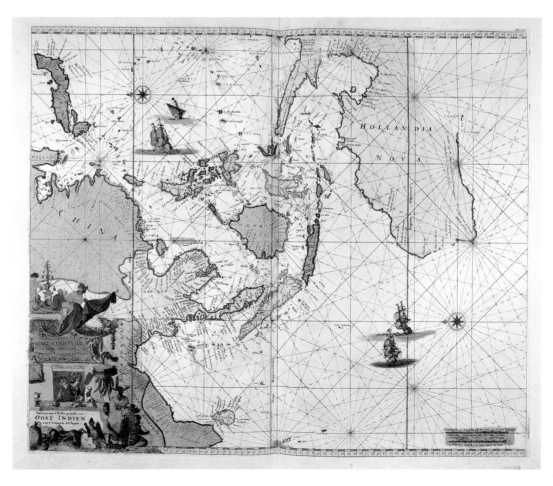

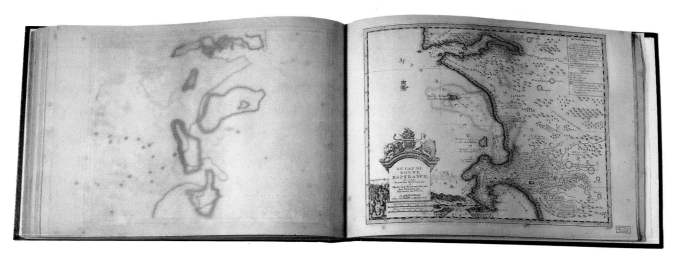

Map of the Cape of Good Hope, South Africa

THE PRINTING AND PUBLISHING COMPANY of Jan Covens and Cornelis Mortier operated in Amsterdam from the early eighteenth century until 1866. The business was established by Pierre Mortier (1661-1711), a publisher and map seller, who issued Sanson's and Jaillot's maps as well as those of de Wit. He acquired de Wit's stock in 1706 and some of the Jansson and Blaeu plates from Schenck. At his death, the company of Covens and Mortier was formed, and plates by Delisle were added to the stock; in 1730 Delisle's maps were reissued as the *Nouvel atlas*.

Among these plates was a detailed map of the Cape of Good Hope, *above*, which includes a vignette of the harbour of Table Bay with the famous flat-topped mountain behind it. In the foreground is a typical Dutch gabled farmhouse, set amid orchards. In 1652 the Dutch East India Company had established a trading station at the foot of Table Mountain, and the Cape had become an essential victualling stop for ships plying between Europe and the Far East. The map indicates that much of the surrounding area, too, had been settled by this date.

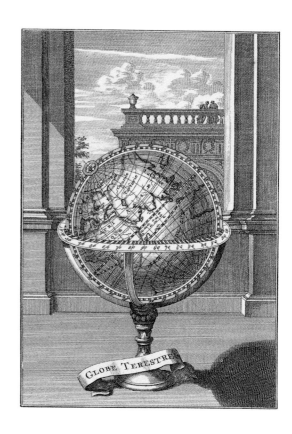

Terrestrial globes were first made at the end of the 15th century, but since they were based on Ptolemy's maps the distribution of the landmasses was misleading, with a vast Terra Australis appearing in the southern hemisphere. During the 17th and 18th centuries maps became more accurate, and celestial and terrestrial globes were made in great number: they were found in almost every private library.

Globes made of wood or metal were covered with printed or manuscript gores, or sections, cut so that they came to a point at the North and South poles. Sometimes circular sections were produced which were cut around the edge so they could be aligned at the Equator. The frame encompassing the globe was marked off in degrees to aid calculation.

This celestial map, with its graphic representations of the constellations, is displayed as two spheres, orientated to suggest that the viewer is standing at the North or South Pole. The northern sphere contains constellations such as the Great Bear and the Twins; Pices and Centaurus can be seen in the southern map, while other constellations, such as Scorpio and Saggitarius, appear across both.

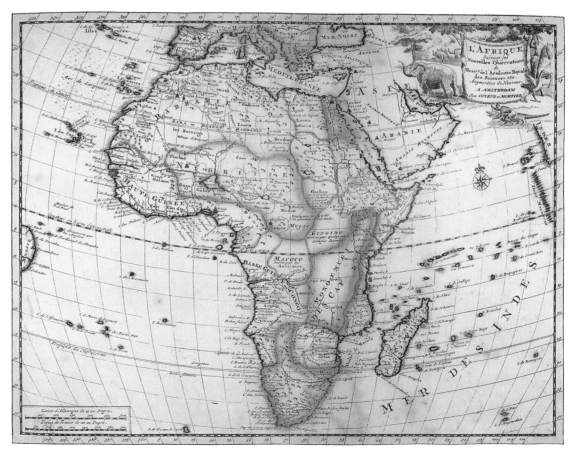

Because this map of Africa is based on the work of Delisle, the Ptolemaic lakes have been deleted and the Nile has been reduced to a more reasonable length. The rivers Gambia and Senegal, emerging from the west coast of the continent, are correctly marked, but the latter's relationship with the Niger is still not understood; the Congo also appears. Interesting features include the Fort of the Hollanders (the beginnings of Cape Town) and, in the north, the trade routes across the Sahara to Timbuktu and elsewhere.

115

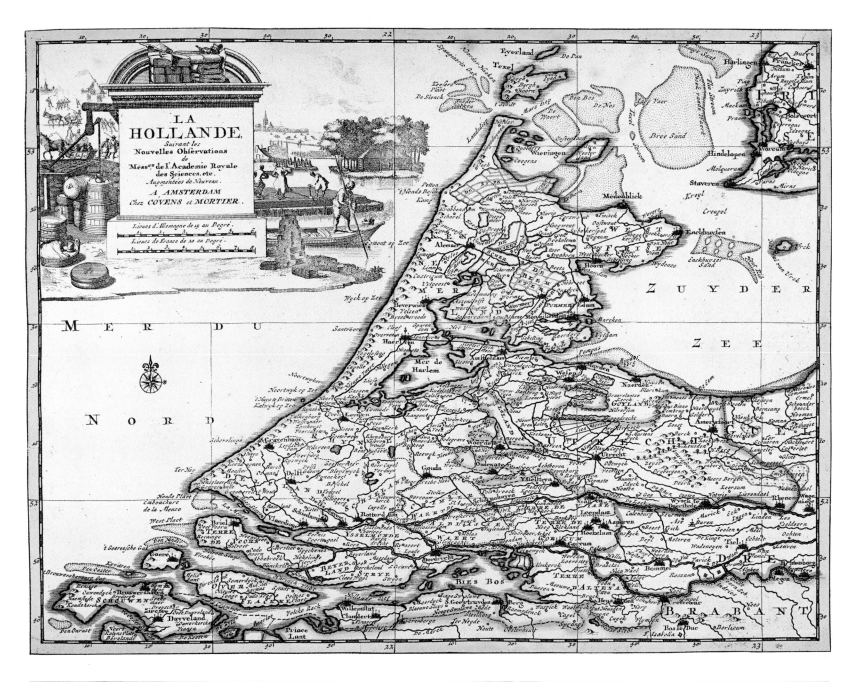

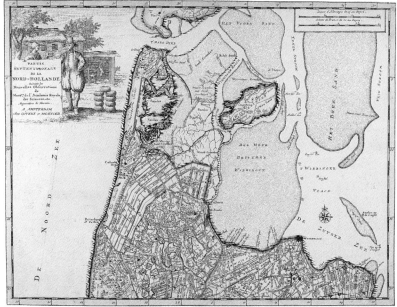

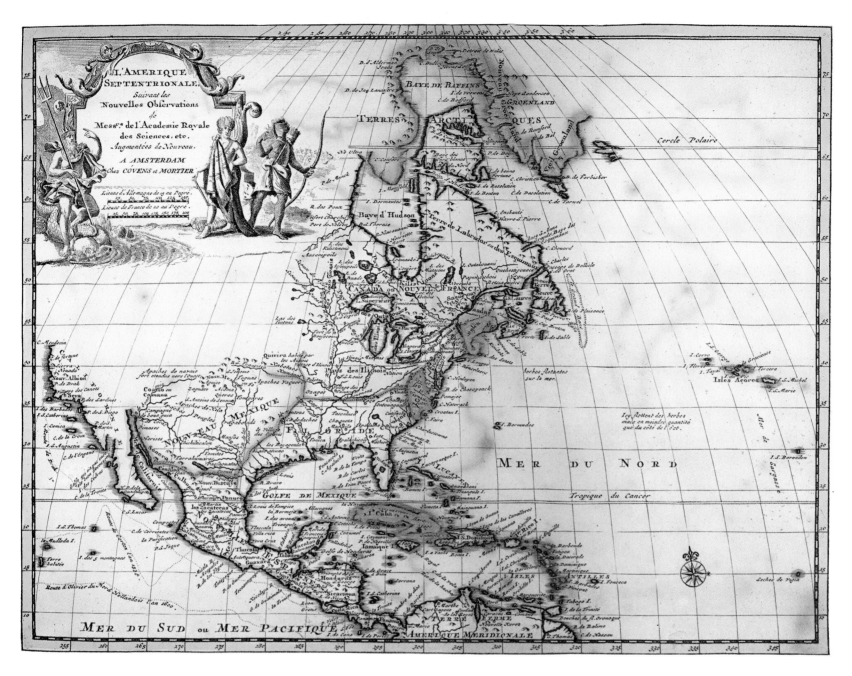

A disproportionate number of maps in the atlas cover Holland, although in varying degrees of detail. The general map of the country, *above left*, has a vignette illustrating aspects of Dutch life: some horse-drawn sledges may be seen on a frozen stretch of water, and a labourer cuts peat for fuel. Some of the implements used to make the cheese crucial to Holland's economy are also shown. The inlet from the Zuiderzee on which Amsterdam is situated here leads into the Sea of Harlem, which has long since been drained.

The map of north Holland, *far left*, shows part of the general map in greater detail. The navigable channel between the Bree Sands and Wieringen Island, which allowed large vessels to reach Amsterdam, is meticulously marked. Here the finished cheeses are shown stacked. Fishing nets and baskets adorn the cartouche to the map of Zeeland, *left*, and Neptune is drawn through the water in a shell chariot by four horses, emphasizing the importance of the sea in this region.

This map of America includes Mexico, parts of South and Central America and the West Indies. It also reaches as far north as Baffin Bay, named after the English Arctic explorer William Baffin (*c*1584–1622). The map title gives distances in both German and French leagues; this measure varied but was usually equivalent to 2½–3 English miles.

California is now correctly drawn as a peninsula, but the west coast is incomplete and most of the interior of the continent has yet to be explored.

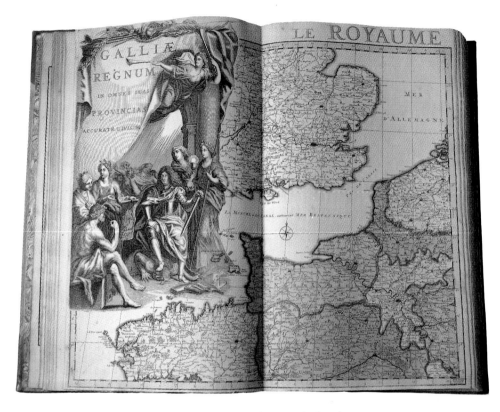

The map of northern France, *right*, is one of several covering the country which claim to show the kingdom "accurately divided into all its provinces."

The king of France is surrounded by figures in classical garb, one of whom holds up a map of Britain, which is symbolized as the footstool of the king.

Map of northern France in the *Nouvel atlas* of 1739

GUILLAUME DELISLE (1675–1726) was the son of a geographer and a pupil of the eminent astronomer Jean Cassini, who came to France from Italy in 1669. Cassini taught Delisle to distrust all "authorities" and to rely upon severe scientific testing. Delisle became a member of the Académie Royale des Sciences in 1702 and was appointed Premier Géographe du Roi in 1718.

In 1739 Delisle's maps were reissued by Covens and Mortier as the *Nouvel atlas*, and although Delisle was credited the elegant cartouches were somewhat coarsened.

During his lifetime, Delisle issued some 60 high-quality maps, in 1700 publishing a world map and in 1700–12 an *Atlas de géographie*; his *Atlas nouveau*, issued in 1730 after his death, ran to many editions, several of which were pirated. His elder brother, Joseph Nicholas, who was also a skilled cartographer, was invited by Peter the Great to teach astronomy at St Petersburg.

With their intimate knowledge of Russia, the Delisles produced the best maps of the country available at the time. That of the Caspian Sea, which is named in seven languages, is orientated with north at the left, as the compass rose indicates. Countries bordering the Caspian, including Russia, Persia, Georgia and Tartary, are indicated, as are important rivers, and there are notes on the map giving information about the nomadic peoples living in the region.

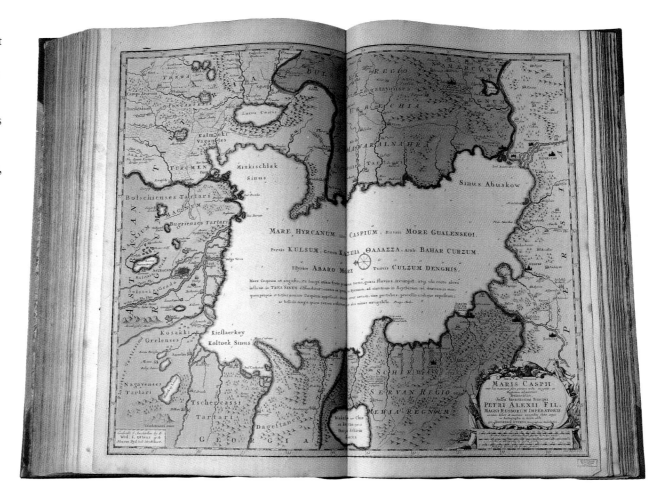

Delisle's maps were redrawn for the 1739 edition by eminent Dutch cartographers—this map of Iceland, for instance, was the work of Pieter Schenck and Gerard Valk.

The map appears to have been made for navigational use, since it contains two compass roses and rhumb lines. The coastline is fairly accurate, especially on the eastern side, but there is little understanding of the interior, and neither the extensive icefields nor the capital, Reykjavik, established c930, is marked.

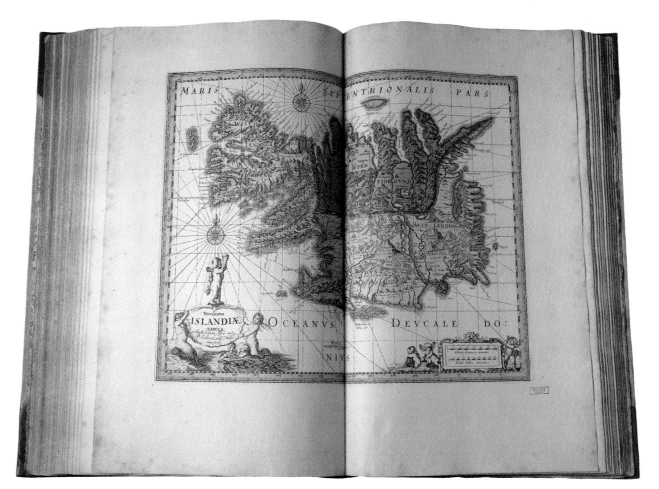

EMMANUEL BOWEN 1766

Cartouche to the map of France

THE LIVES OF EMANUEL BOWEN (*c*1700–67) and his son Thomas (*fl*1749–90), both engravers and publishers, spanned the century during which the French map makers were overtaken by the British, and Emanuel became Engraver of Maps to both King George II of England and Louis XV of France. Despite this, he died almost penniless.

Thomas engraved plates for his father's atlas and produced individual maps which appeared in some of the most important English topographical works of the eighteenth century, such as Morant's *Essex* (1768) and James Cook's *Voyages* (1773). Financially, however, he fared even worse than his father, dying a pauper in Clerkenwell workhouse.

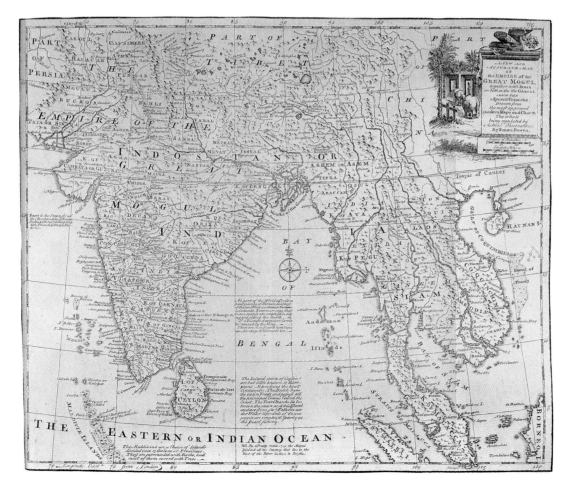

The cartouche to the map of France, *above*, claims that it has been composed from the most modern maps and charts and regulated by astronomical observation. Distances are measured in English and French leagues, "20 to a degree".

Emanuel Bowen's "new and accurate map of the Empire of the Great Mogul" appeared in *The Maps and Charts to the Modern Part of the Universal History* (1766). A note close to Ceylon (Sri Lanka) states that it produces the best cinnamon; another note identifies India as a source of diamonds, with the rider that "If a diamond exceeds 10 carets 'tis claimed by the King."

The volcano of Pic d'Orixba, viewed from the Forest of Xalapa

B ARON ALEXANDER VON HUMBOLDT, a distinguished German naturalist, went to South America in 1799 with Aimé Bonpland. He spent the next five years exploring Colombia, Venezuela, Ecuador, Peru, Cuba and Mexico.

Humboldt published an account of his travels in 1807. This was followed in 1812 by his atlas of "New Spain" (Mexico) with maps with much scientific and topographical detail and information about natural features.

Acapulco, on the west coast of Mexico, was founded in 1550. Its deep-water harbour served as a haven for Spaniards exploring the Pacific and the city later played a crucial role in trade with the Philippines. Today it is largely a holiday resort, despite suffering frequent earthquakes and hurricanes.

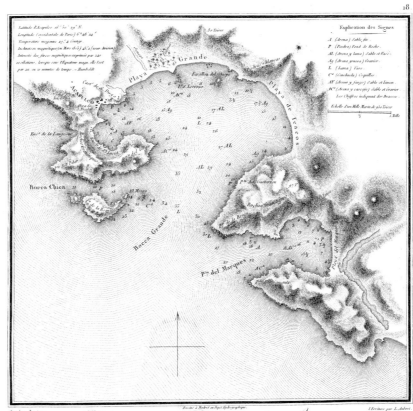

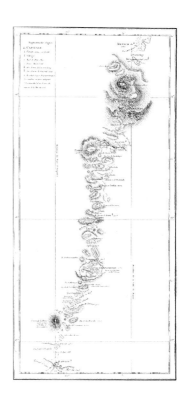

This detail covers the area from the valley on the high plateau where Mexico City lies to Acapulco. The survey was made by Miguel Costanzo and Garcia Conde, following Humboldt's observations.

MATTHEW FLINDERS 1814

North coast of Port Bowen at 7am on 24 August 1802

MATTHEW FLINDERS (1774–1815), the English hydrographer, navigator and explorer, entered the Royal Navy in 1789 and went as a midshipman on the *Reliance* to New South Wales in 1795. He spent much time with the ship's surgeon, George Bass, recording the Australian coast and explored the George River and the unknown area south of Port Jackson. In 1801 he returned to Australia with a party of scientists aboard the *Investigator*.

Flinders undertook a thorough survey of the coast and became the first man to circumnavigate the continent. On his return voyage, the condition of his ship forced him to stop at Mauritius, where he was held prisoner by the French for six years. His health suffered in consequence, but during this time he wrote many scientific works, including *A Voyage to Terra Australis*, accompanied by a volume of maps, of which these are examples.

In 1642 Abel Tasman discovered Van Diemen's Land (Tasmania) and named it after the governor-general of the Dutch East Indies, Anthony van Diemen. Captain James Cook, sighting it on his 1772 voyage, thought it was part of the mainland, but Flinders and Bass proved it was an island by sailing around it.

Port Lincoln on Spencer Gulf appears as an inset, with that of Nuyts' Land, on a general map of Australia's south coast.

The inset, *below*, from a map of south Australia, shows Nuyts' Archipelago, named after Pieter Nuyts, a high official in the Dutch East India Company. He was on board the ship *Gulden Zeepard* which sailed eastward along the coast in 1627 as far as St Peter and St Francis islands.

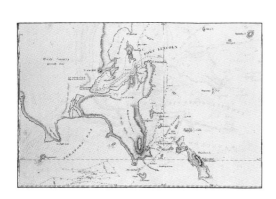

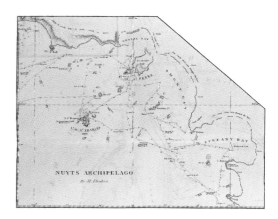

H ENRY S. TANNER'S *New American Atlas*, with 22 maps—5 of the world and the continents and 17 of the American states—was published in 1823 at Philadelphia, then the centre of American map making. Around 1815, the states' economy and population began to grow rapidly: there was a cotton boom in the Deep South and increasing farm settlement on the plains north of the Ohio River.

Title page, with vignette of the first
landing by Columbus in the New World

By the Louisiana Purchase (1803), the States acquired from France the formerly Spanish region of Louisiana, then stretching far to the north. At the Anglo-American Convention of 1818, the 49th parallel was agreed as the boundary between the United States and Canada, and in 1819 Spain ceded Florida and renounced her claim to Oregon. Tanner's maps reflect these great developments.

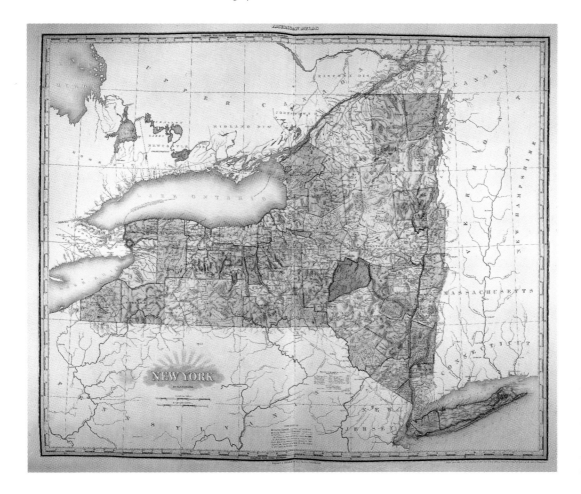

Tanner's map of New York State, sub-divided into its counties, is meticulously executed, for by that time the east coast, although not the interior, had been thoroughly charted and distances were accurately known. The Great Lakes and parts of the region's numerous river systems are indicated.

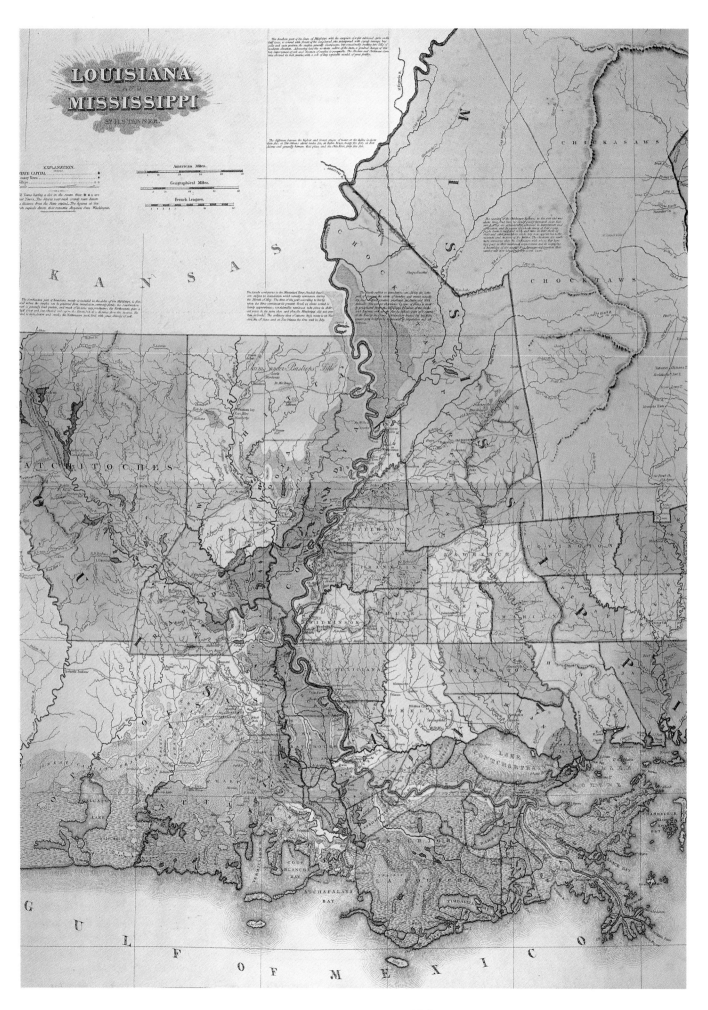

The map of Louisiana, the eighteenth state to be admitted to the Union, and Mississippi, the twentieth, reflects their importance at the time as the foremost cotton-producing areas in America. The advent of steamboats on the Mississippi (the first navigated the river to New Orleans in 1812) greatly enhanced the region's prosperity.

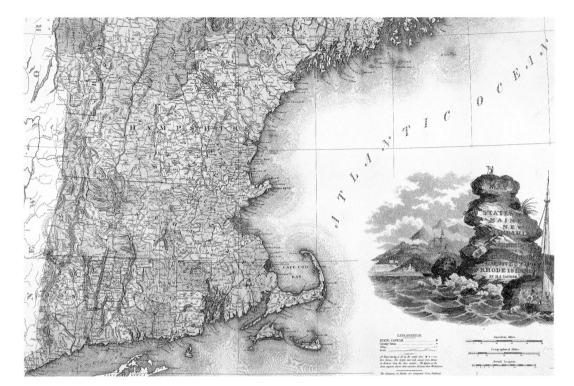

The title cartouche, *right*, to Tanner's map of Maine, New Hampshire, Vermont, Massachusetts, Connecticut and Rhode Island emphasizes the importance of the fishing industry on the east coast at the time. At the top of the rocks, a lone figure views the mainland through a telescope.

The coast and northern regions of Africa are well documented on the map, *below*, but a vast area in the south is marked simply "unknown parts". The river systems of the interior are also largely excluded, since they were not successfully explored by Europeans until the middle of the century. But in the year the atlas was published, 1823, the English explorer Hugh Clapperton (1788–1827) reached Lake Chad.

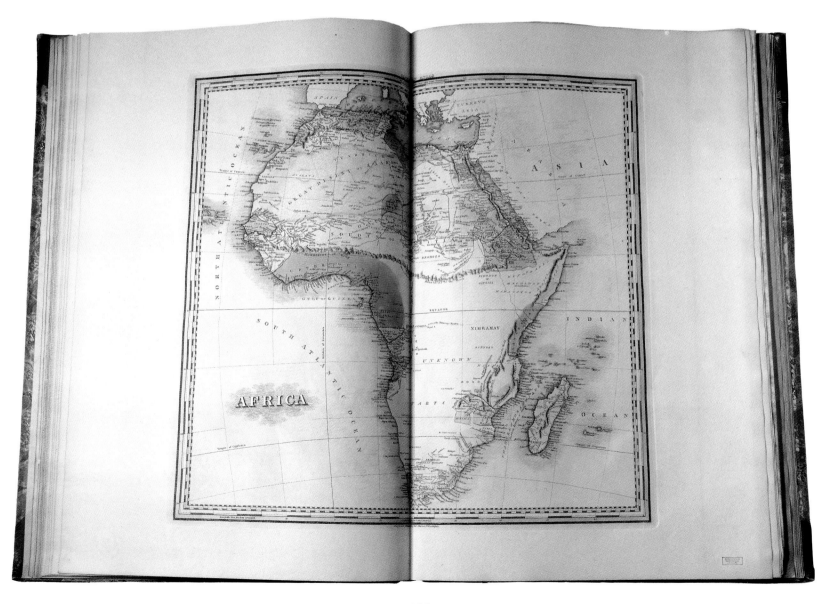

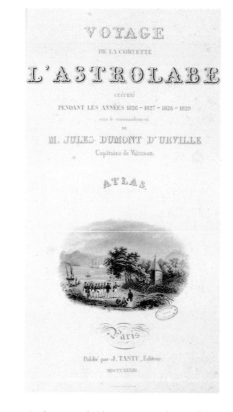

JULES-SÉBASTIEN-CÉSAR DUMONT D'URVILLE, born in France in 1790, is renowned as a navigator who commanded two great expeditions. On the first, to the South Pacific (1826-29) aboard the *Astrolabe*, he discovered or redesignated several island groups, instigating the revision of many existing maps and charts. He returned to France with more than 1,000 plant specimens, 900 rock samples and much information about the inhabitants of the islands he had visited.

Steel-engraved title page to a volume of the atlas recording the voyage of the *Astrolabe*

On Dumont D'Urville's second expedition, to the Antarctic, he hoped to sail beyond the most southerly point reached by James Weddell in 1823, but his ships were ill-equipped to penetrate the pack ice. He did, however, discover Joinville Island and Louis Philippe Land and, sailing into the Antarctic again on his return voyage across the Pacific, he sighted the Adelie coast south of Australia, naming it for his wife. He and his family died in a railway accident in France in 1842.

Dumont D'Urville's voyage to the South Pacific was meticulously documented, and in 1833 a six-volume atlas was published in Paris. As well as detailed charts, these contained information about the flora, fauna and peoples he had observed and topographical and geographical data.

The fishes, *left*, are taken from Volume II. At bottom left is *Holocentre lion*, one of a family of fish with narrow compressed bodies. It was found off Vanikoro, the others, off islands as far west as New Guinea. The birds on the plate from Volume III were sighted on Vanikoro, New Guinea and Guam.

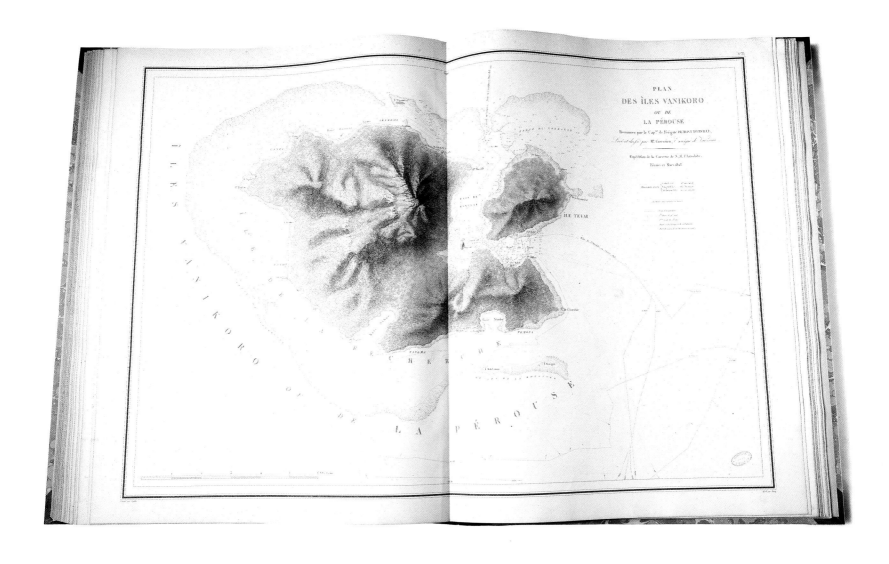

These landmarks, all drawn in 1827 and published in Volume VI, are located on the coast of New Zealand. The island of Otea, *top*, was recorded on 20 February at a distance of 4mls/6.5km; the Cape of Otou, *centre*, was drawn on 7 March from 10mls/16km, and the isle of Kandabon, *bottom*, on 3 June from 3mls/5km.

The islands of Vanikoro, part of the Santa Cruz group, were visited by the *Astrolabe* in February and March 1828. The ship's route and the soundings taken are shown on this plate of the main island of Nendo, dominated by a volcano, which Alvaro de Mendaña the Spanish navigator attempted to colonize in 1595.

D'Urville believed that wrecks sighted here were the ships of the explorer La Pérouse, who had disappeared in 1788.

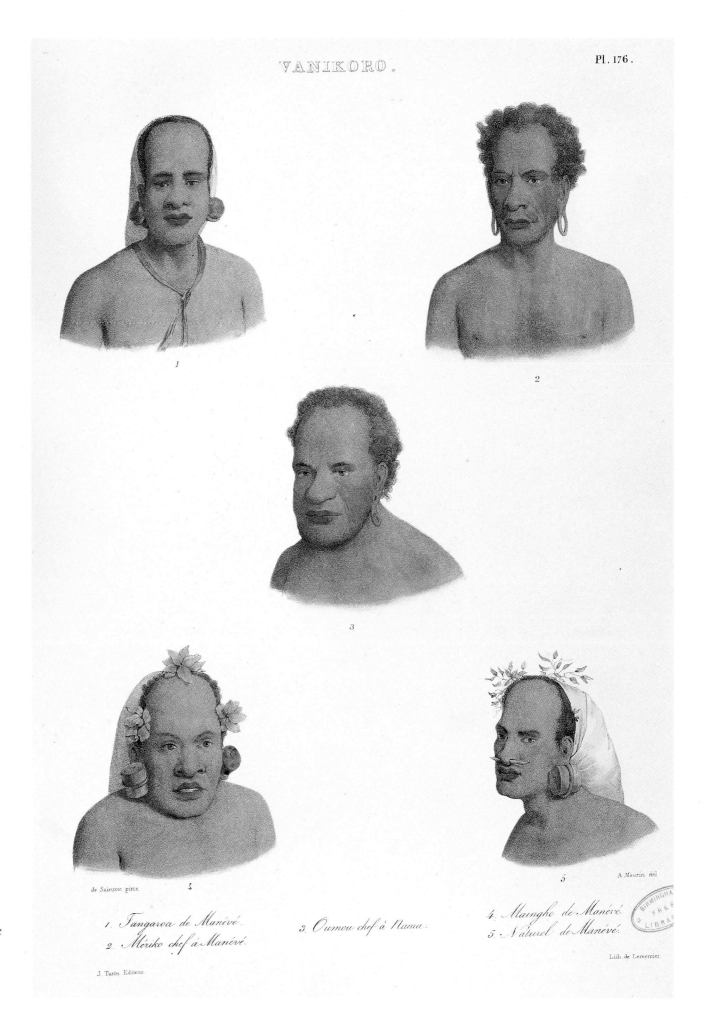

VANIKORO.

Pl. 176.

1

2

3

de Sainson pinx

4

A. Maurin del.

5

1. *Tangaroa de Manévé.*
2. *Mériko chef à Manévé.*

3. *Oumou chef à Nama.*

4. *Maingho de Manévé.*
5. *Naturel de Manévé.*

J. Tastu Éditeur.

Lith. de Lemercier.

The artists accompanying Dumont D'Urville were fascinated by the appearance of the different peoples they encountered on the Pacific islands, and many plates are devoted to picturing them.

The portraits here are of chiefs and men of importance on Vanikoro. Three of the men wear large ornaments in their ears; when these were removed the distended lobes hung slackly, as the ears of the other two show. The two at the bottom also wear head-dresses held in place with flowers and leaves.

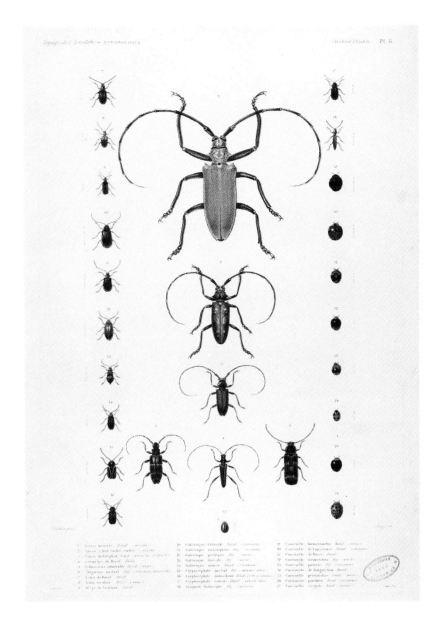

Plate 8 in the entomological section of the atlas shows a wide variety of coleopterans, or beetles, ranging from the tropical goliath beetle to tiny round ladybirds with shiny, often spotted, wingcases.

All the beetles are identified by number in the list below.

Differences in the appearance of the various islanders is carefully noted: the men from Vanikoro, *right*, have short, woolly hair and nose and ear ornaments, while those from Tikopia have long hair hiding their ears, and tattooing on face, chest and back.

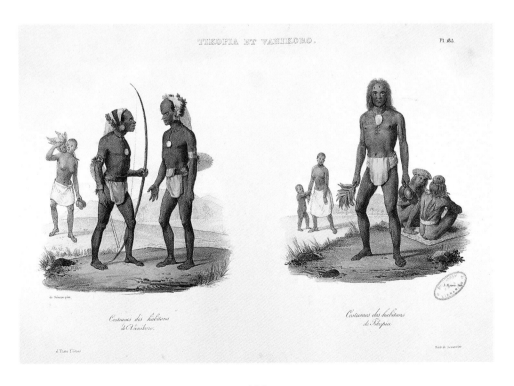

The first butterfly on this plate shows the large female of the species *Ornithoptera priam* found in the Celebes, with a smaller view of the butterfly's underside. The third butterfly is *Danaide cousine*, one of the monarch butterflies, from Vanikoro.

JOHN ROCQUE 1746

JOHN ROCQUE, probably the son of Huguenots who fled from France after the Revocation of the Edict of Nantes in 1685, was born c1704. He was, possibly, trained as a surveyor: his first recorded work was a survey of Richmond House in 1734. By that time he also had a shop in Piccadilly in which he sold prints.

In 1737 he and the engraver John Pine published a map of London, for which there was a great demand since the city had been rebuilt and was growing rapidly following its virtual destruction in the Great Fire of 1666. The map was published in 1744–46 in two editions: the full 24-sheet version at eight guineas and single sheets at 5s each.

Dedication to Sir Richard Hoare, Lord Mayor of London, and the Aldermen

This section of the River Thames, *below*, traversing London from the new Westminster Bridge to just beyond the Tower of London, covers four sheets of Rocque's map. Features that may be readily identified include, west to east, the Houses of Parliament, the open space of Covent Garden Market, Somerset House on the Strand and the River Fleet, once the boundary of the City of London.

St Paul's Cathedral, recently designed by Sir Christopher Wren, and old London Bridge, still with houses on it, are clearly shown. The great number of ships lying on the river below the bridge confirms London's importance as a port.

The atlas was meant to be of practical use and included an alphabetical index of streets.

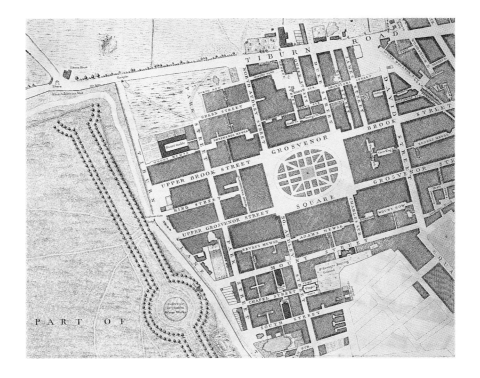

A detail of the Hyde Park area shows Tiburn Way (present-day Park Lane) running between Hyde Park Corner and Tiburn Road (today known as Oxford Street). Tiburn gallows, where public executions took place until 1783, are also marked.

A new, fashionable residential area east of Tiburn Way and centred on Grosvenor Square was being developed on the estate of the Grosvenor family, later the Dukes of Westminster. The name Mayfair, given to the district, derives from an annual fair, much dreaded by the residents, who found those that attended it a "throng of beggarly, sluttish strumpets".

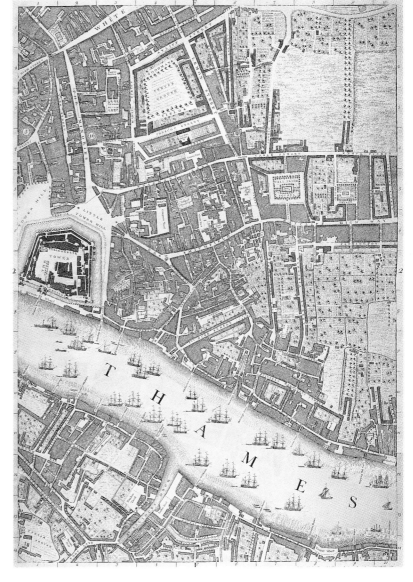

TAYLOR AND SKINNER 1776

As the section, *below*, of the Edinburgh–Inverness route shows, the roads were arranged in strips, to be read left to right and bottom to top. When a page had been used, it was turned, and then the atlas itself by 180°, to follow the rest of the journey.

THE BRITISH TRADITION of the detailed road atlas, established with John Ogilby's *Britannia* in 1675, was continued by the surveyors George Taylor and John Skinner. Their publications included *The Roads of North Britain and Scotland* (1776) and *The Roads of Ireland* (1777).

In the early eighteenth century, cartography was dominated by engravers who simply copied each other's maps. But as the century progressed, advances in mathematics and astronomy, initiated by Isaac Newton, produced professional surveyors, whose work revolutionized map making.

Part of the route from Edinburgh to Stirling and Fort William appears below. Distances are indicated at intervals of a mile and important landmarks are shown, among them, of crucial importance to travellers, many hostelries.

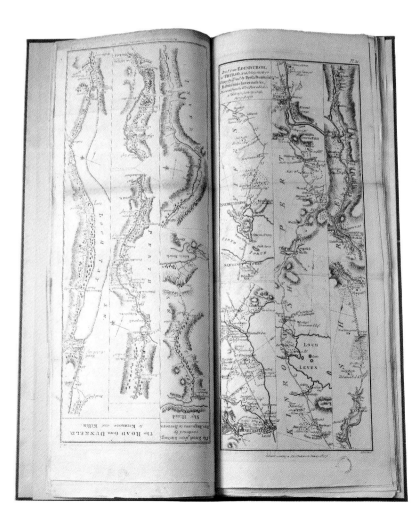

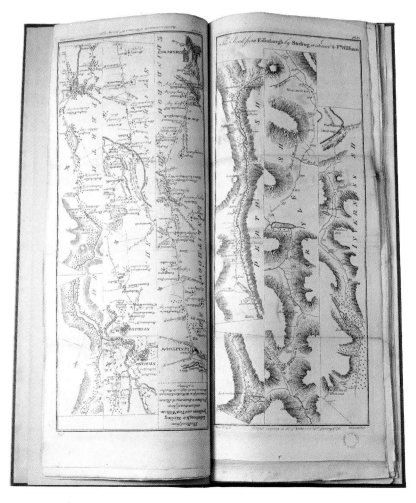

JOHN CARY 1809

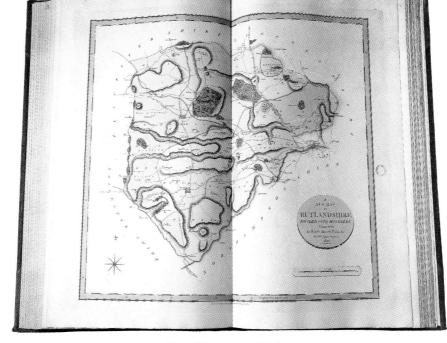

JOHN CARY, who was an eminent English cartographer, was born in about 1754. After serving an apprenticeship as an engraver, in 1783 he established himself in the Strand, London, and soon gained a high reputation as an engraver and globe maker.

In 1794 he was commissioned by the Postmaster General to survey the roads of England; the resulting *English Atlas* of 1797 was followed by many further editions. Cary also made maps for the Ordnance Survey until 1805, when a separate establishment was set up.

Map of the county of Rutland

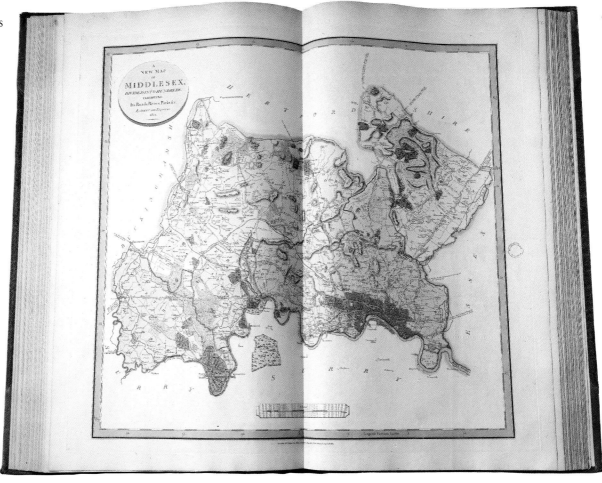

Cary's cartography was of a very high order, as the maps from the 1809 edition show. That, *above*, dated 1801, is of Rutland, the smallest of the old English counties.

The map of Middlesex, *left*, may be compared with Norden's map of 1593, where the cities of Westminster and London had only recently merged to form an urban area.

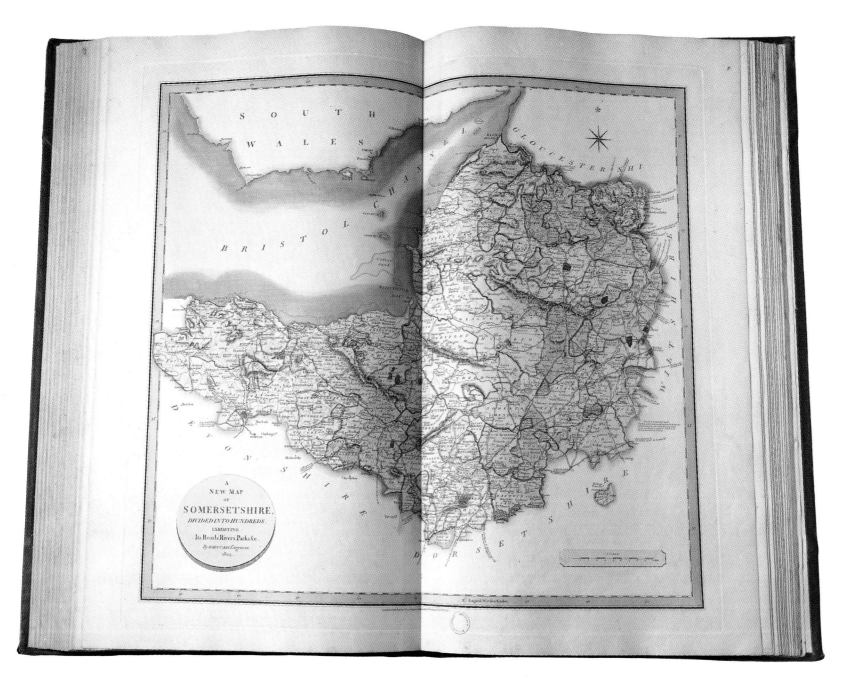

As was Cary's custom, the map of the English county of Somerset was divided into its Hundreds. It is generally thought that originally a Hundred consisted of 100 geld hides, the basic Anglo-Saxon land unit for the purpose of taxation, but Hundreds varied greatly from county to county both in number and size. They had their own judicial courts but slowly lost their importance and from the 16th century had only a formal existence.

The map of Cornwall shows its famous tin and copper mines, known to ancient Greek and Phoenician traders, which were reopened during World War II. Remote from the rest of the country, Cornwall still has its own local language, related to the Welsh and Breton tongues.

The Scilly Isles, a rural district of the county of Cornwall, appear as an inset on the Cornwall map, *left*. They lie 28mls/45km off the coast and comprise 150 islands, five of which are inhabited.

Title page to the second part of the *Carte générale*

BACLER DALBE (1761–1824), an engineer and geographer, was director of the Topographical Bureau of the War in Italy, established during the French revolutionary period. The *Carte générale du théâtre de la guerre en Italie et dans les Alpes* was published in Milan between 1798 and 1802, after the city fell to Napoleon Bonaparte in 1796.

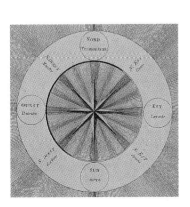

The elaborate title page, *above*, of the second part of Dalbe's *Carte générale* includes a cartouche displaying imperial symbols: wreaths of laurel, cannon, pikes, spears and banners. The neo-classicism of the period was used to bolster Napoleon's imperial ambitions, and the vignette depicts the French "liberating" the Italian people from Austrian rule.

The second part of Dalbe's atlas contains maps of the "Kingdom of Naples, Sicily, Sardinia and Malta", here shown on a single plate. The numbered divisions identify individual plates, each of which covers a specific theatre of war; this complete section of the atlas comprises 24 plates.

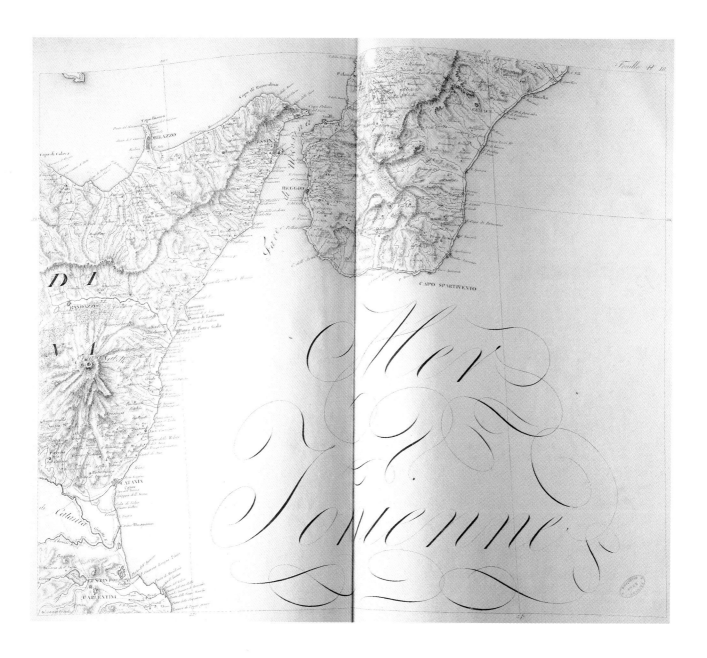

The dominant feature on Plate 18 of the Italian maps, showing the northeast corner of Sicily and the toe of Italy, is the volcano of Mount Etna. The strategic ports of Messina and Regio, lying either side of the Strait of Messina, are marked, as is the headland of Scylla. From ancient times, the passage between this and the whirlpool Charybdis filled mariners with trepidation.

The standard of typography and engraving throughout Dalbe's atlas is superb. The maps themselves are engraved in meticulous detail, while the militarist and imperialist decorations to the title pages are executed with great artistry and skill.

CARTE TOPOGRAPHIQUE DE L'EGYPT 1818

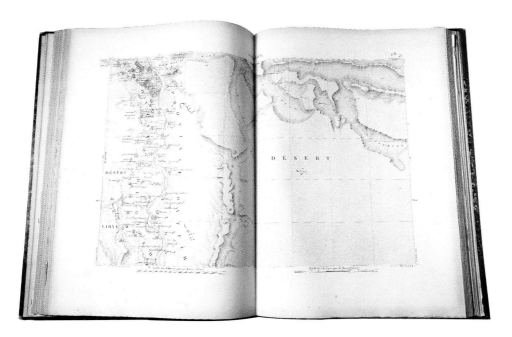

Map of the area around Memphis, Egypt

THE *CARTE TOPOGRAPHIQUE de l'Egypte*, edited by Jacotin and published in Paris in 1808, was compiled after Napoleon's campaign in Egypt in 1798 by military engineers and geographers. The scale is large, 1mm:100m.

The map, *above*, of the area around Memphis includes the pyramids at Giza, outside Cairo, and other topographical details such as fields irrigated from the Nile and man-made canals. Place names are given in both French and Arabic.

This is a detail from a map of the Rosetta (moden Rashid) and Burlos area on the north coast of Egypt.

Rosetta, which is spelt Rosette on the maps, lies on one of several distributaries of the Nile. Salt flats lie between the desert and the sea, and ancient ruins are indicated.

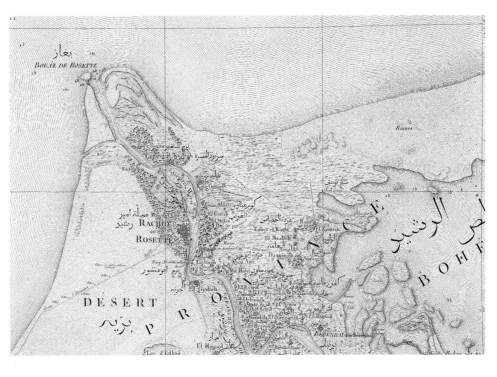

Near the town of Rosetta, lying within an area of palm groves, the famous Rosetta Stone was found in 1799 by a Frenchman. Since it carried inscriptions in both Greek and hieroglyphs, it provided a key to the eventual deciphering of Egyptian hieroglyphic writing by Jean-François Champollion and Thomas Young.

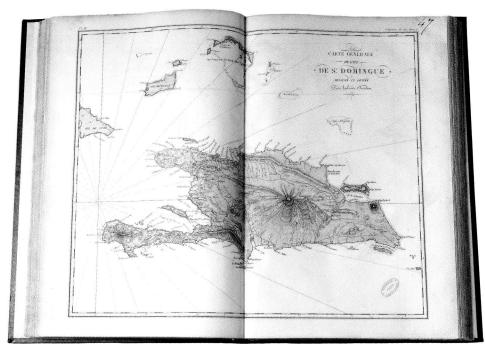

The island of St Dominique, where Columbus landed in 1492, remained in Spanish hands until 1844.

THE FRENCH ENGRAVER and cartographer Ambrose Tardieu was born in 1788. He worked both for general publishers and for the Depôt de la Marine, and in 1821 contributed maps to the *Histoire générale des voyages de Jean François de la Harpe*, the French explorer.

Tardieu also produced maps for the *Nouvel atlas* of Langlois and, in 1833, for Dumont D'Urville's *Voyage de la corvette L'Astrolabe*. His final work, which was published in 1846, the year of his death, was the *Carte routière d'Italie*.

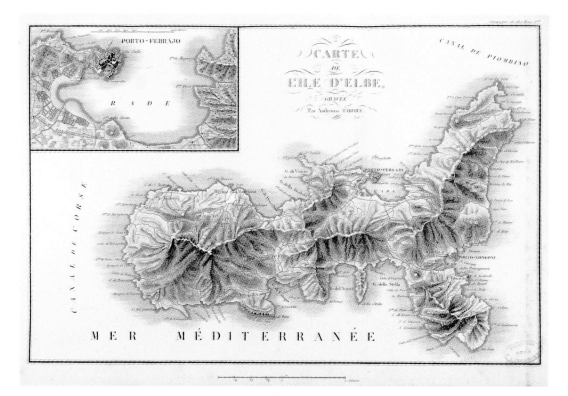

The mountainous island of Elba, lying some 6mls/8.5km from the Italian mainland, forms part of the province of Livorno. It was to Elba that Napoleon was banished in 1814, after his enforced abdication, and from which he returned for the "100 days" that ended at the Battle of Waterloo. The inset shows the capital, Portoferraío, on the island's north coast.

This plan of the citadel and town of Alessandria in Piedmont was drawn in 1822, after Napoleon's death. It shows the elaborate defences built by the French in 1803–l4. The 11th-century fortress, erected by the Lombard League against Frederick Barbarossa, was strengthened to form a base for the French advance across the north Italian plain.

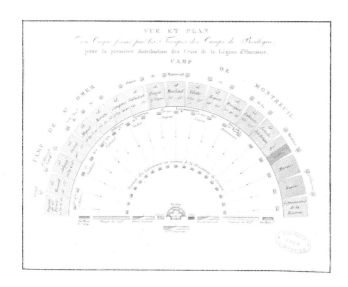

The view, *left*, of Napoleon's camp at Boulogne in 1805, with his army paraded for a distribution of the Croix de la Légion d'Honneur, is accompanied by a diagram, *right*, identifying the various formations. The recipients of the award form a semicircle in front of Napoleon's throne at right centre. Officers, also in a semicircle, stand in front of the infantry units, behind which the cavalry is drawn up.

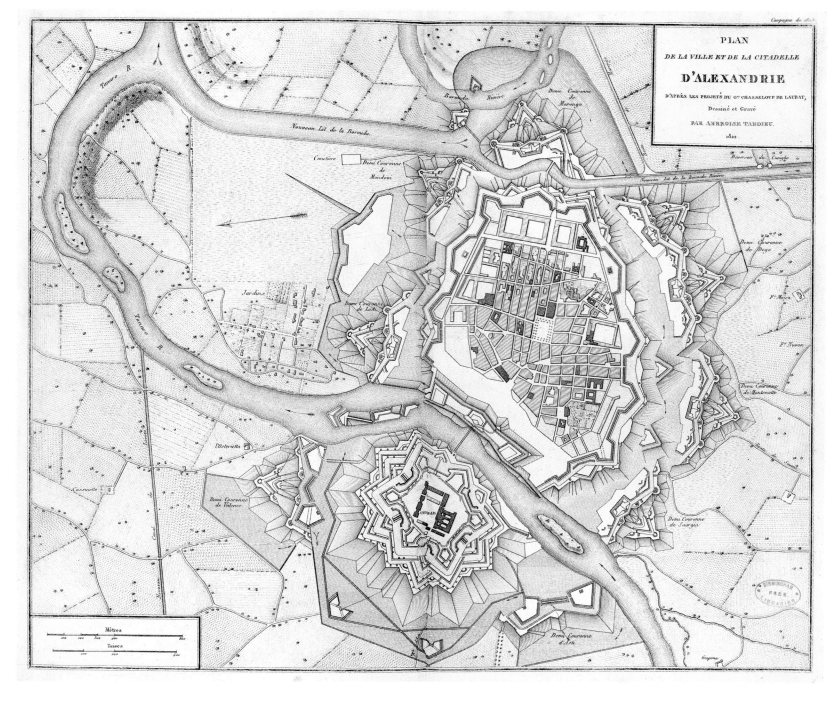

"The Times" Atlas.

Containing 118 Pages of Maps, and comprising 175 Maps

and an Alphabetical Index

to 130,000 Names.

PUBLISHED AT

THE OFFICE OF "THE TIMES."

PRINTING HOUSE SQUARE, LONDON, E.C.

1897.

5-81

CHAPTER FIVE

The
MODERN AGE

THE SECOND HALF of the eighteenth century had been a period of extraordinary cartographic activity throughout Europe and especially so in Britain. This was largely because of the scientific advances in the fields of trigonometry and astronomy, which were initiated by Newton, and the vital innovations in instrument design made by Hadley, Harrison and Ramsden, all of whom were Englishmen.

John Hadley (1682–1744), who was an optician by trade, built a reflecting telescope, the first of sufficient accuracy and power to be of use in astronomy. He also invented a quadrant that would evolve into the sextant. John Harrison (1693–1776) was a watchmaker who invented a marine chronometer, which enabled mariners accurately to calculate their longitude at sea, while Jesse Ramsden (1735–1800) was an outstanding pioneeer in the design of precision instruments, including an extremely accurate sextant and theodolite, and barometers.

The advent of the Industrial Revolution created much wealth and allowed Britain further to expand its colonies, which soon became markets for mass-produced British goods. The new lands also became

Part of a world map in two hemispheres which appears in Blackie's *Imperial Atlas*, published in London in 1860.

The simplicity of the typeface on the title page of *The Times Atlas* of 1897 is in stark contrast with the vignette of the clockface, in the decorative style favoured at the period.

a focus for exploration and the information gained, combined with scientific and technical advances in cartography, allowed it to assume an important place in the cartographic world toward the end of the nineteenth century.

It was, however, in the field of sheet-map production that Britain gained pre-eminence, and this was stimulated by awards of up to £100, made between 1759 and 1809 by the Society of Arts, to any person who completed a county survey on a scale of 1in:1ml/ 2.5cm:1.6km. By the early nineteenth century almost the whole country had been mapped at this scale and the Ordnance Survey, established in 1791-92, adopted the scale and continued the work.

The Revolutionary and Napoleonic wars had a great impact on cartography both in France, where Napoleon's campaigns created a demand for accuracy and detail, and in countries such as Britain which wished to know about territory that might need defending or that it or its allies might be required to attack. The nineteenth century saw also the completion of large-scale charting of continental coasts and much exploration in the continents of Africa and Australia.

It was in Australia that Matthew Flinders completed the work first undertaken by the Dutch and continued by Captain James Cook. This and similar national surveys were conducted on lines established in the late 1600s by the Cassinis, a family of renowned astronomers; the accuracy of the surveys was increased by recent improvements in the design of instruments and by greater scientific knowledge.

Printing techniques also changed in the mid-nineteenth century, and the copper plates of earlier times were replaced by lithography, a process whereby the image was drawn on stone. This allowed for both the taking of thousands of printed copies without loss of clarity and the introduction of a mass of detail and colour: earlier copper plates had rapidly became worn and frequently had to be recut to retain clarity. Copperplate engraving did not entirely disappear, however, until the invention of photography and its application to lithographic printing.

The idea of a general atlas comprising maps of all parts of the world had existed since the days of Mercator and had been exploited by the great Dutch publishers Blaeu and Jansson, but in the nineteenth century the wealth of scientific information available allowed cartographers to include thematic and other types of map in their atlases. These atlases thus became "expressions" of the geographical thinking of particular countries and the foundation in Germany of geography as a subject for study contributed greatly to this trend.

Baron Alexander von Humboldt's *Atlas géographique et physique,* published in 1812 to accompany his studies in New Spain (Mexico), includes a vast amount of information rendered in cartographic form, and he had plans to produce another atlas

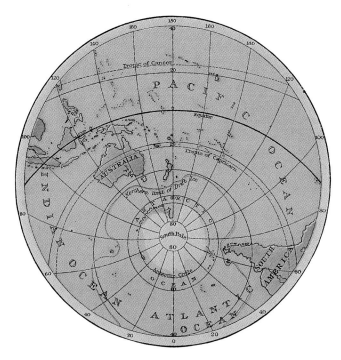

HEMISPHERE CONTAINING GREATEST AMOUNT OF WATER.

The inset map of the southern hemisphere in *The Times Atlas* dramatically illustrates the massive proportion of water to land. The Pacific Ocean, named by Magellan, alone occupies almost one-third of the surface of the earth.

Tahiti, the largest island in the Windward Group of the Society Islands, was a French protectorate in 1860, when this map was published in Blackie's atlas. The island comprises two eroded volcanic cones, Tahiti Nui and Tahiti Iti, which are joined by the Isthmus of Taravao.

to illustrate geographic concepts. Humboldt's ideas were soon taken up by his compatriots Adolf Steiler and Heinrich Berghaus and, in France, by Denaix. In 1838 the first edition of the *Physikalischer Atlas* of Berghaus was published, to be followed in 1852 by a revised and extended version which included information about meteorology, climatology, hydrology and geology, as well as rainfall and the distribution of races and even of some diseases.

The importance of Berghaus's atlas was appreciated in Britain, and the Edinburgh publisher, Alexander Keith Johnston, included some of the maps in his *National Atlas* of 1843. When his *Physical Atlas* was issued two years later, two-thirds of the maps were derived from Berghaus. Rivals of Keith Johnston were the two John Bartholomews, father and son, whose company established the Edinburgh Geographical Institute. At the very end of the nineteenth century, they proposed to publish an impressive *Physical Atlas* in five volumes. Sadly, only the tomes covering zoogeography and metereology were ever published.

The production of general atlases was dominated by German commercial publishers until the end of the century, although the French school of geography saw its ideas expressed in the *Atlas géneral Vidal-Lablache,* which was published in 1894. The first edition of *The Times Atlas,* produced in 1895 was, in reality, an English-language version of Andreo's *Allgemeine Hand Atlas*; but after two further editions it was completely redesigned by John George Bartholomew to reflect the territorial changes after World War I.

The portable general atlases that became such a feature of the era reflect the interests and preoccupations of both the publishers and the readers, and British publications

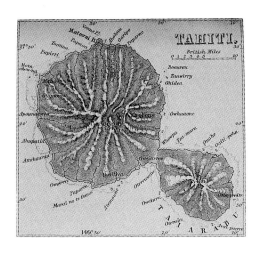

emphasized the extent of the empire. They were also keen to show the colonial possessions of the other European powers in order to record the relatively greater size of British territory.

Although the eighteenth was the century of rapid colonial growth, for the European powers the nineteenth was largely the era of consolidation. Until the mid-1800s, the old colonies of Spain and Portugal had remained intact, and those of Britain had grown, but one great area was still largely unexploited: Africa. The lion's share of Africa fell to Great Britain, closely followed by France, with Spain, Portugal, Germany, Italy, Belgium, and even Turkey, trailing along behind.

Until about 1800, few Europeans had penetrated to the interior, but as colonization developed so explorers—Mungo Park, David Livingstone and James Bruce among them—probed deeper, providing detailed information for cartographers. By 1900, the continent had been so effectively carved up that only Morocco, Ethiopia and Liberia remained independent sovereign states; and even Morocco was a French protectorate. The redistribution of European colonial possessions following World War I and their break-up after World War II, however, lie outside the scope of the atlas collection under discussion.

The Spanish possessions in both North and South America and in the Philippines, which had been gained in the sixteenth century, remained intact until the nineteenth century, but the Peninsular

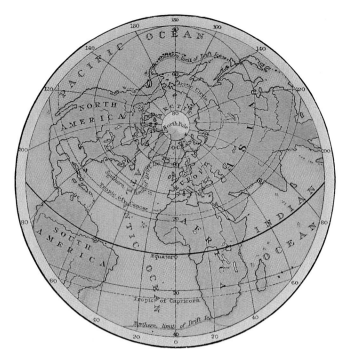

HEMISPHERE CONTAINING GREATEST AMOUNT OF LAND.

The northern hemisphere contains relatively little water and the bulk of the world's landmasses, as this companion to the map, *opposite*, shows. Paradoxically, the North Pole is ice-covered water, the South Pole, ice-covered land.

War of 1807-14 left the country weak, and the restored monarch Ferdinand VII, sought to re-establish the absolutism of the previous century. It was during his reign, which lasted until 1833, that a series of revolts in the Americas denied Spain most of its colonial possessions except those in the West Indies and in the East. The Spanish–American War of 1898 ended, however, with the evacuation of Cuba and the Philippines, and the ceding of Puerto Rica to the United States.

The keynotes of the nineteenth century, and particularly of the Victorian age in Britain, were energy, self-confidence and a firm belief in progress. Industrial development, which had begun in Britain, spread to other countries in Europe and to America and resulted in increased productivity, while medical advances ensured an expanding population that demanded better education and services.

The development of a railway network, a postal system and the telegraph went hand in hand with the growth of popular newspapers, the public library system and better public health and led to the formation of a society which required books that both informed and entertained. The introductions to many of the general atlases published toward the end of the century expressed the hope that they would fulfill this function—and most of them certainly do so.

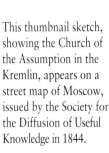

This thumbnail sketch, showing the Church of the Assumption in the Kremlin, appears on a street map of Moscow, issued by the Society for the Diffusion of Useful Knowledge in 1844.

SOCIETY FOR THE DIFFUSION OF USEFUL KNOWLEDGE 1844

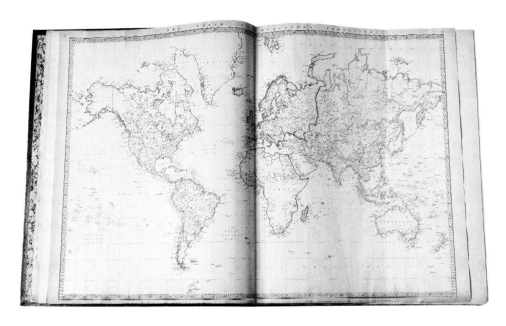

Map of the world in *Maps of the Society for the Diffusion of Useful Knowledge*, 1844

The site of Pompeii, buried by the eruption of Vesuvius in AD 79, was discovered in 1748 and by 1844 sporadic excavation had uncovered a great part of it.

A map of the town appeared in this atlas with, beneath it, an artist's panorama of the main landmarks. Naples is shown at the far left, and Vesuvius itself in the middle distance.

T HE SOCIETY for the Diffusion of Useful Knowledge was founded in England in 1826 to provide improving information at affordable prices for people who would otherwise have been deprived of education.

Not all the published titles were of equal clarity of language, merit or scholarship, but in its early days the Society was highly successful because many writers and scholars were prepared to work for little or no remuneration. This goes some way toward explaining the enormous body of work it was able to publish over a period of 20 years.

The Society soon appreciated that cheap, accurate maps were needed to supplement readers' studies, for such as existed were often difficult to obtain and always expensive. The series continued from 1829 to 1843, and surviving sets of these maps are now highly priced.

Eventually the Society faced financial problems. Its *Biographical Dictionary*, published in half volumes, had achieved only seven, covering the letter "A", when work ceased because, it was reported, "many persons, who fully approve of the plan and execution, decline to take the work until it is complete."

The sale of maps was among sources of income that never failed, but despite this the Society was obliged to suspend operations in March 1846.

A plate of the world's main rivers, *right*, depicts them all draining into a central lake—a common device at this time. The rivers of Africa, America, Asia and Europe are identified by different types of wavy line. A key, not shown in this detail, lists the rivers in alphabetical order, gives the length of each in English miles and assigns a number to it. The numbers are repeated in a circle within the lake, enabling the details of any river to be checked with ease.

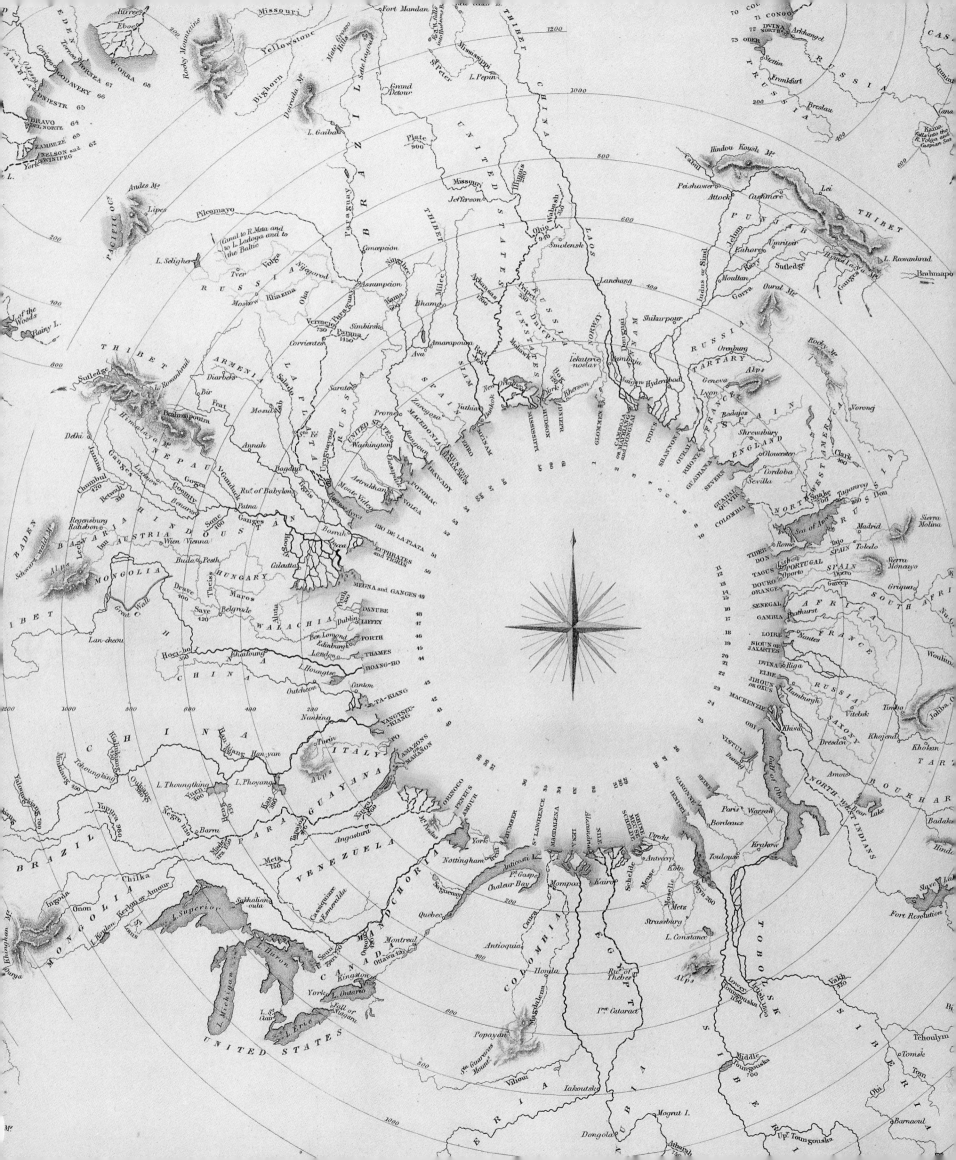

NEW YORK.

SCALE OF FEET

1000 500 0 1000 2000 2640 feet = ½ mile

JERSEY CITY

NEW JERSEY RAILROAD

POWLES HOOK

ELLIS OR BUCKING ISLAND

Fort Gibson

HUDSON RIVER

NEW YORK

HARBOUR

Castle William

GOVERNORS OR NUTTEN ISLAND

FORT COLUMBUS

S. Battery

Castle Clinton or Gardens

U.S. Revenue Office

Whitehall St.

BATTERY

Bowling Green

BROADWAY

College Pl.

Chapel Street

St. Paul's

Park

WASHINGTON SQUARE

BROADWAY

FAYETTE PLACE

BOWERY

Stuyvesant Pl.

TOMPKINS SQUARE

AVENUE

AVENUE

AVENUE

FIR

BUTTERMILK CHANNEL

EAST RIVER OR STRAIT

Corners Mill Pond

Furman

Columbia

State

Joralemon

Hicks

Henry

Clinton

Court

LONG ISLAND RAILROAD

Sidney Pl.

Livingston

Schermerhorn

Pacific

Dean

Bergen

Wyckoff

Butler

Baltic

Sackett

Union

President

Congress

Strong Pl.

CITY OF BROOKLYN

Monroe

STREET

Clinton

MAIN STREET

FULTON STREET

Moore

Prospect

Navy Yard Ferry

Lit. Dock

WALLABOUT BAY

CORLEARS HOOK

Williamsburgh Ferry

WILLIAMSBU

Second

Third

Fourth

Fifth

River Street

Grand Street

By 1844, maps of Japan were becoming extremely accurate, despite the fact that Japan had been closed to foreigners for around 200 years.

The insets at the top of the map show, on the left, Nagasaki, its harbour and environs, and Yedo (Tokyo). This city, in 1844, had a population of 1,200,000 and was the official residence of the shogun, or military ruler.

The red lines around some of the islands are peculiar to this copy of the atlas but were probably added at the time of publication.

In the early days of New York's growth, the main roads ran west to east, linking the trading activities on the Hudson and East rivers. Much later, further development gave added importance to the north–south avenues. The grid pattern was imposed in 1811 when the city was first laid out; only Broadway, which followed a centuries-old track of the Weckquaesgeck Indians, retained its diagonal path across Manhattan Island. The red lines indicate the New Jersey Turnpike and the Harlem and Long Island railroads, both built in 1832. Curiously, in the list of Principal Buildings, No 12 is not identified.

Top to bottom: the Church of St Charles Borromeo; St Steven's Dom Kirche (the cathedral); Church of the Jesuits, *left*, and the Church in Turri.

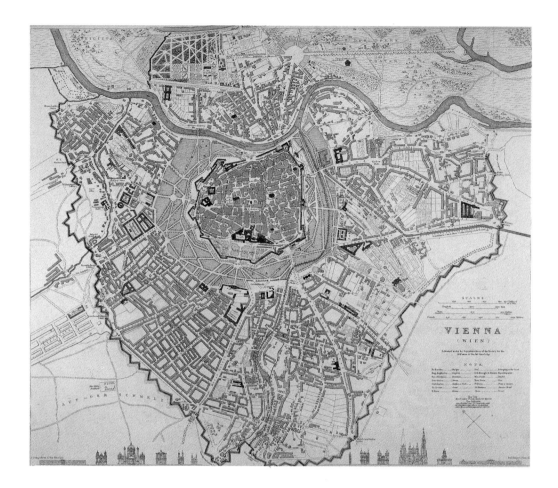

The city of Vienna, founded in Roman times as a frontier-fortress against Germanic tribes, lies in a strong position on the right bank of the Danube, which here divides into several arms. It is also the hub of an ancient north–south trade route.

The old city, with its 13th-century walls, is girt with woods and meadows; but long before 1844, buildings had spread beyond this to the outer wall, erected in 1704–06 and dismantled in 1890. By 1845, the population of Vienna was 430,000.

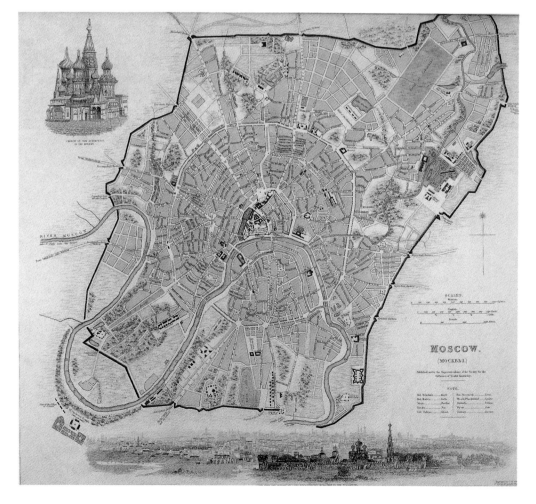

This detailed map of Moscow, astride its river, includes an artist's impression of the city with the Novo Devitchei monastery in the foreground. In the centre of the plan, outlined in black, is the fortified enclosure of the Kremlin, the heart of the city. As with most old cities, Moscow developed in concentric rings around this inner core, with roads radiating out like the spokes of a wheel.

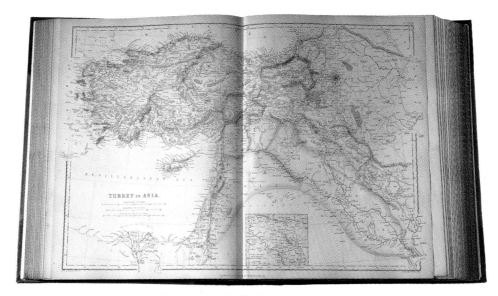

Map showing the Turkish Empire in Asia

I N 1860, *The Imperial Atlas of Modern Geography* was published by Blackie & Son. It was conceived and edited under the direction of Dr Walter Graham Blackie, a Fellow of the Royal Geographical Society in London and an enthusiastic educationalist; Blackie intended his atlas to be "useful and acceptable to the general reader".

His introduction states that "special prominence is given in this work to Great Britain and her colonies" and, indeed, 19 images, a quarter of the whole, are devoted to this theme. He pays tribute to many individual geographers and cartographers, among them John Bartholomew junior, who was entrusted with drawing and engraving some of the maps.

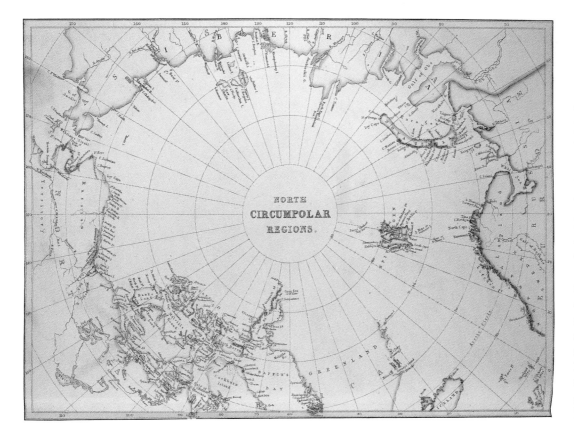

At the time of the map, *above*, Turkey's Asian possessions still included present-day Israel, Iraq, Jordan and Syria.

At the height of its power, at the end of the 16th century, most of the Middle East formed part of the Ottoman Empire, but over the centuries, as a result of wars with Persia, Egypt and Russia, Turkey lost much of its territory in Asia. Losses in Europe contributed to the empire's decline, and the whole edifice collapsed with Turkey's defeat at the end of World War I.

The map of the North Pole and its environs, *left*, was made before the flattening of the earth at its poles had been precisely measured.

Sir Isaac Newton, the English mathematician and natural philosopher, calculated early in the 18th century that the earth must be flattened at its poles, but he could not be certain by what amount. Ever since his estimate, a reliable figure had been sought: with the recent advent of satellites it was found to be 26.6mls/42.8km.

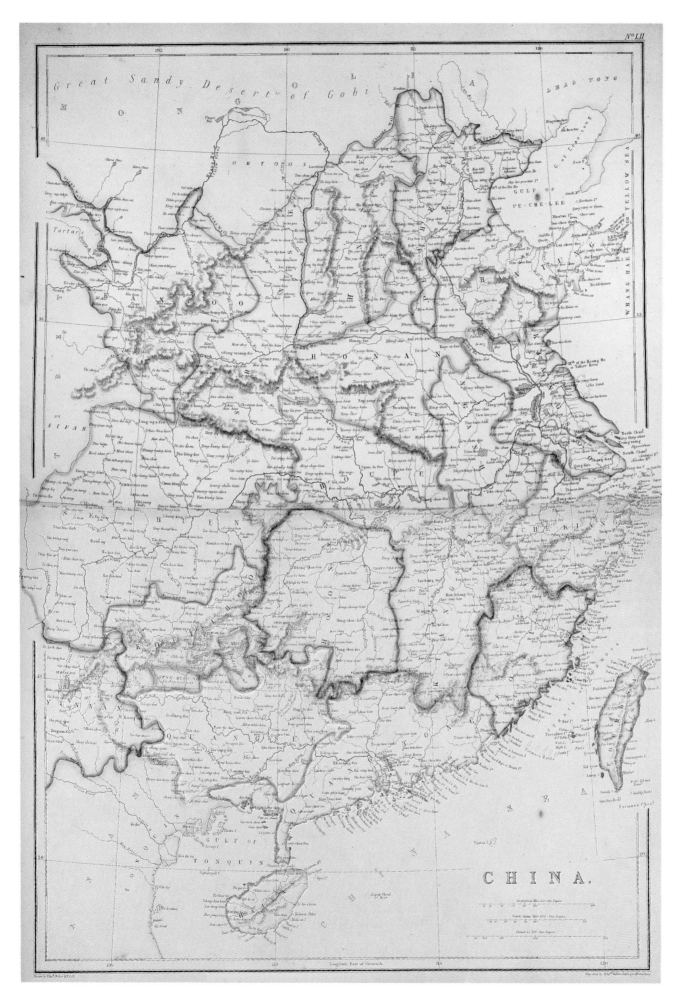

The continents and their adjacent islands are outlined in different colours in the map of the world, *right*. Thus, in the western hemisphere, New Zealand, incorrectly drawn as a single island, is coloured blue like Australia, and the Russian Empire is clearly divided into Europe and Asia by the Urals.

The map of the Chinese Empire, *left*, shows it at its greatest extent, reached under the emperors Kangxi (*r*1662–1722) and Quianlung (*r*1735–96) of the Qing Dynasty (1644–1912).

The disintegration of this vast empire in the second half of the 19th century was of considerable interest in Great Britain, for Hong Kong had been ceded to her by China by the Treaty of Nanking in 1842.

In the map of the Caucasus, *right*, orange is used to outline the Turkish Empire and blue to indicate Persian territory, on the right. All the rest are provinces of the Russian Empire; those outlined in green are independent states within it, and Georgia is coloured pink.

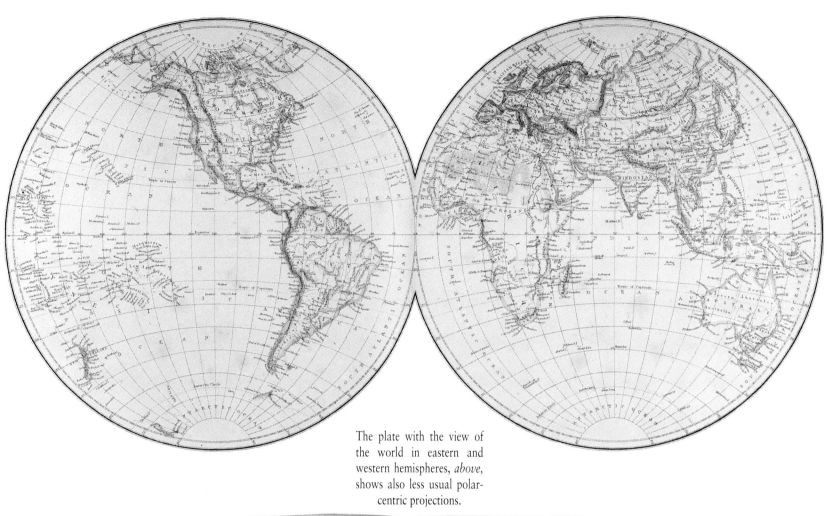

The plate with the view of the world in eastern and western hemispheres, *above*, shows also less usual polar-centric projections.

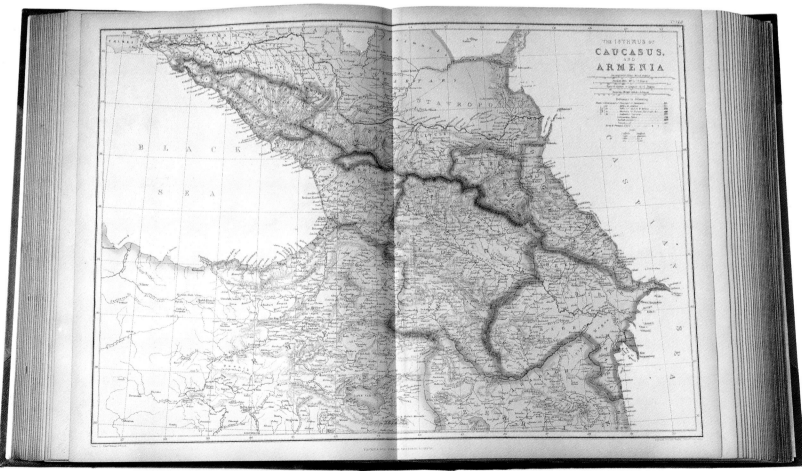

J. G. BARTHOLOMEW 1892

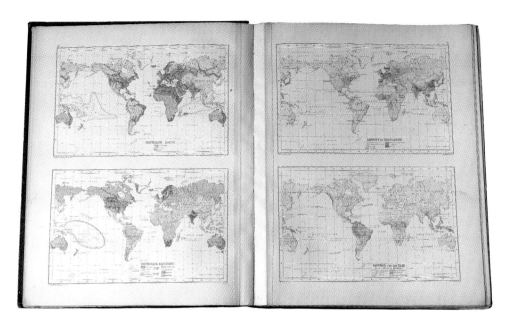

Maps showing the races, populations, religions and rainfall of the world

JOHN BARTHOLOMEW (1805–61), founder of the Edinburgh firm of map makers, was an engraver who produced a range of maps for guide books and schools. His son, also John, who trained as a draughtsman and engraver, expanded the business after his father's death.

He worked mainly in orthodox cartography but, notably, contributed the frontispiece map for Robert Louis Stevenson's *Treasure Island*, first published in 1883. His most important innovation, in the late 1870s, was the introduction of contour layer colouring to indicate different heights of land.

The plate, *right*, from the 1892 edition of the *English Imperial Atlas* deals with astronomical geography. The phases of the moon and the ocean tides, the solar system and the comparative size of the planets are graphically demonstrated.

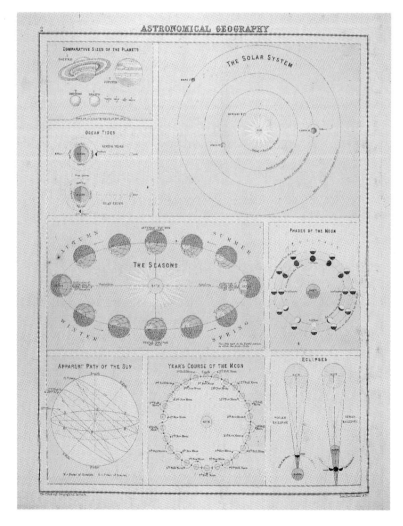

The grandson of the founder, John George Bartholomew (1860–1920), who took over the firm in 1888, introduced many technical improvements in map making. One of his first acts was the compilation of *The English Imperial Atlas and Gazetteer of the World*, which was published in London in 1892 by Thomas Nelson. This was one of several works issued by publishers using maps by the Bartholomew family of cartographers.

John George was also responsible for merging the business with that of Thomas Nelson, thus forming the Edinburgh Geographical Institute.

The sheer bulk of information offered, on a wide variety of subjects, reflected the determination of John G. Bartholomew to make his atlases appeal to the growing number of literate people in all walks of life who were now able to afford them.

The colophon of the Edinburgh Geographical Institute, as it appears in the 1892 *Imperial Atlas*.

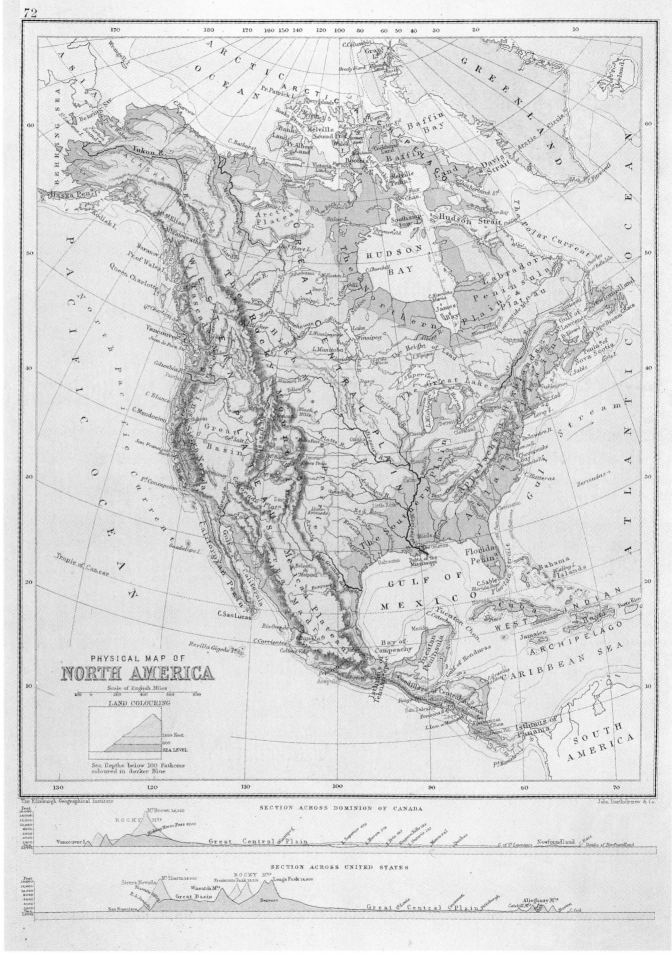

In this map of North America, Bartholomew used the system of layered colouring. Because this was relatively new, the map is also equipped with cross-sections and a key to indicate the land heights represented by the colours. The sea is shown in a uniform blue, but the layered colouring system was soon to be introduced to show the depth of water also.

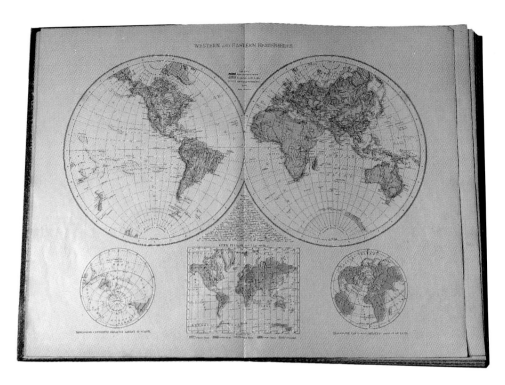

Plate with 60 of the world's highest mountains marked and numbered

T*HE TIMES*, a London newspaper, in 1895 resolved to produce a major new atlas for Great Britain. German cartographers, regarded as the most gifted practitioners of the art at the time, were entrusted with drawing the 118 maps.

As with many atlases of the period, this contained a plate explaining graphically the solar system and the movements of the planets and their satellites. Although some 18th-century atlases had included these features, they became commonplace only in the 1800s because of the wider educational function atlases were then expected to provide.

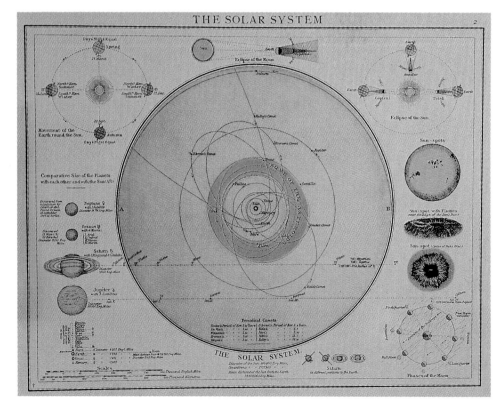

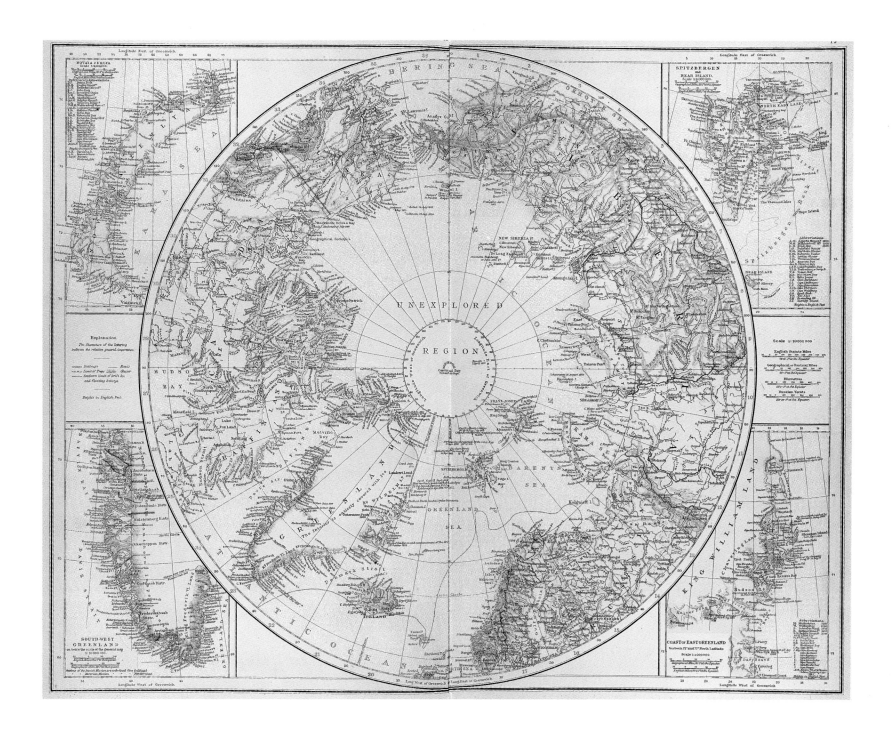

The 18th and 19th centuries produced an unprecedented flow of information about regions hitherto unknown to Europeans and the first seven plates in this atlas show the physical and human geography of the entire world. But half the plates in this first edition, which was often reprinted, are still devoted to Europe.

When a second edition was needed to record new information and the political changes after World War I, *The Times* entrusted the cartography and printing to the firm of Bartholomew, with which it collaborates on its atlases to this day. The new atlas, *The Times Survey Atlas of the World*, comprised 112 double-page maps, but now only a third were of Europe. In the important five-volume Mid-Century Edition that appeared in 1955-59, following World War II, Europe occupied even fewer pages as the colonial era came to an end.

This plate of the north polar region includes the landmasses of Europe, Asia and the Americas and some features that had only recently been established, such as the vast icefield in the interior of Greenland. However, King Oscar Land is also marked, despite having been proved not to exist in 1893.

This device, in the Arts and Crafts style of the period, appears on the title page and alludes to "Times past and times future", with a clock signifying time's remorseless passage.

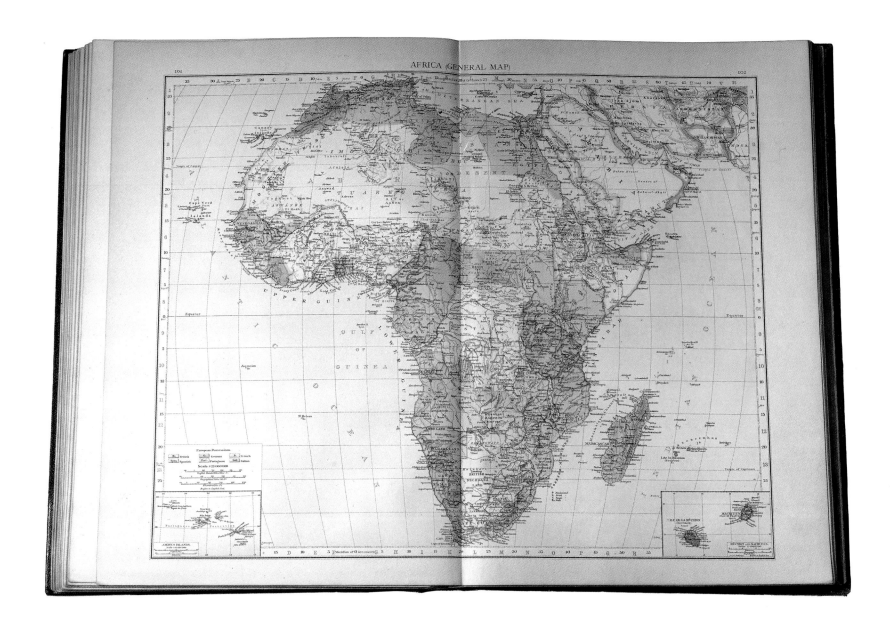

By the last decade of the 19th century, the European powers—Great Britain, France, Germany, Portugal and Italy—had taken possession of the greater part of Africa. An inset gives the colour code used to denote each country's colonies.

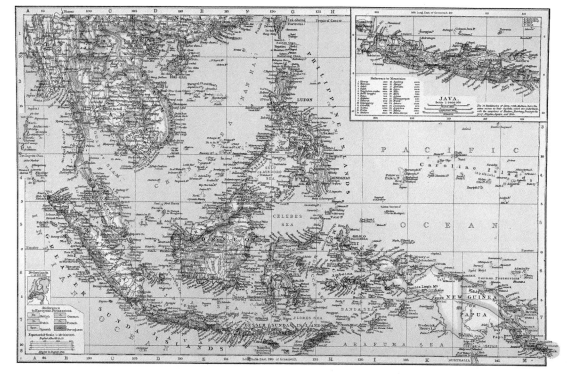

The map of Siam and the Malay Archipelago includes those areas whose valued spices had long lured European explorers and merchants. The Straits of Malacca, charted since the 16th century, which give access to the South China Sea, are dominated by British imperial possessions, notably Singapore. Insets include Java, a colour code and a tiny same-scale map of Holland, which still had extensive possessions in the East.

156

Maps of the moon were an innovation made possible by recent advances in astronomy and by the introduction of photography. This enhanced the telescope's power to penetrate space, for by long exposure of a photographic plate images could be built up that were too dim for the eye to detect.

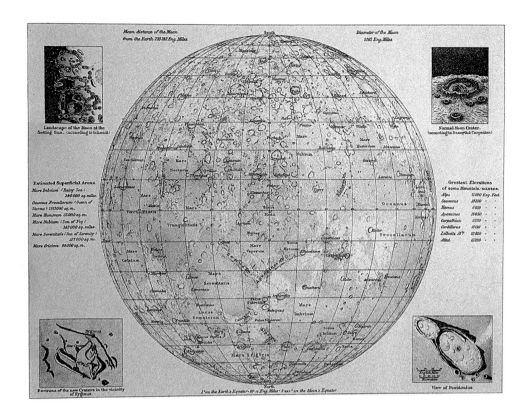

The 48 Spanish provinces and the 17 administrative districts of Portugal are named but not shown on the map of the Iberian Peninsula.

Curiously, the names of the old divisions of Spain, which had been abolished in 1835, are also given, as are the Spanish colonies in North Africa.

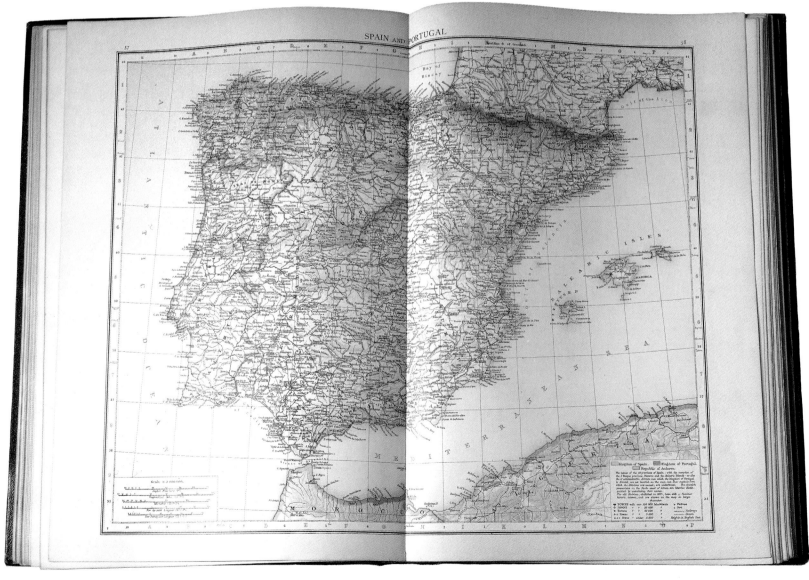

157

INDEX

INDEX OF MAPS

LIST of ATLASES

All the atlases in the Cadbury Collection dating from before 1800 are included in this list; after that, only the atlases discussed in this book are listed.

Ptolemy *Cosmographia* Ulm: L. Holle, 1482
Ptolemy *Cosmographia* Rome: Petrus de Turre, 1490
Ptolemy *Geographiae* Rome: Bernardinus Venetus de Vitalibus, 1508
Ptolemy *Liber Geographiae* Venice: Jacobus Pentius de Leucho, 1511
Ptolemy *Geographiae opus* Strassburg: J. Schott, 1513
Ptolemy *Geographia* Strassburg: J. Schott, 1520
Ptolemy *Opus geographiae* Strassburg: J. Grieninger, 1522
Ptolemy *Geographicae* Lyons: M. and G. Trechsel, 1535
Ptolemy *Geographia* Basel: Henricus Petrus, 1541
Ptolemy *Geographia* Basel: Henricus Petrus, 1550
Ptolemy *Geographia* Venice: Vincentius Valgrisius, 1562
Abraham Ortelius *Theatrum orbis terrarum* Antwerp: Aegid. Coppenius Diesth, 1570
Georg Braun and Frans Hogenberg *Civitates Orbis Terrarum* Cologne, 1572–1618
Gerard de Jode *Speculum orbis terrarum* Antwerp, 1578
Christopher Saxton *Atlas of the counties of England and Wales* London, 1579
A. Lafreri and C. Duchetti *Geografia* Rome, c1580–85
Lucas Janssen Waghenaer *The Mariners Mirrour* London: J Charlwood, 1588
Gerard de Jode and Cornelis de Jode *Speculum orbis terrarum* Antwerp, 1593
John Norden *Speculum Britanniae* London, 1593
John Norden *Speculum Britanniae* London, 1598
Jan Huygen van Linschoten *His Discours of Voyages into ye Easte & West Indies* London: John Wolfe, 1598
Ptolemy *Geografia* Venice: G. Battista and G. Galignani Fratelli, 1598
Abraham Ortelius *Theatrum orbis terrarum* Antwerp: J. B. Vrintius, 1603
Gerardus Mercator *Atlas Minor* Amsterdam: Jodocus Hondius, 1610
Gerardus Mercator and Jodocus Hondius *Atlas* Amsterdam: J. Hondius, 1613
John Speed *The Theatre of the Empire of Great Britaine* London, 1614
Gerardus Mercator and Jodocus Hondius *Atlas* Amsterdam: H. Hondius, 1623
Gerardus Mercator and Jodocus Hondius *Atlas* Amsterdam: H. Hondius, 1633
Gerardus Mercator and Jodocus Hondius *Atlas* Amsterdam: H. Hondius and J. Jansson, 1636
Jacob van Langweren *A direction for the English Traviller* London, 1643
Jan Jansson *Nouvel atlas* Amsterdam, 1650–56
John Speed *England, Wales, Scotland and Ireland described* London, 1662
Hendrick Doncker *De Zee-Atlas* Amsterdam, 1663
John Speed *A Prospect of the World* London: MS for Roger Rea, 1665
Jan Blaeu *Le Grand Atlas ou cosmographie Blaeuiane* Amsterdam, 1667
Pieter Goos *De Zee Atlas* Amsterdam, 1668
Frederick de Wit (and others) *Atlas* Amsterdam, 1671
John Ogilby *Britannia* London, 1675
William Petty *Hiberniae delineatio* London, 1685
Charles Pené (and others) *Le Neptune françois* Paris: H. Jaillot, 1693

Romein de Hooge *Cartes Marines a l'usage des armées du Roy de la Grande Bretagne* Amsterdam: P. Mortier, 1693
N. Sanson *Atlas novum* Amsterdam: P. Mortier, c1700
Justus Danckerts *Atlas* Amsterdam, c1710
Louis Renard *Atlas de la navigation et du commerce* Amsterdam, 1715
J. D. Kohler *Atlas scholasticus et itinerarius/Schul-und Reisen Atlas* Nuremberg, 1718
Thomas Gardner *A pocket guide to the English traveller* London, 1719
Nouvel Atlas Amsterdam: Covens & Mortier, c1720
J. B. B. d'Anville *Atlas géneral* Paris, 1730–90
Hermann Moll *Atlas minor* London: T. Bowles and J. Bowles, 1732
Guillaume Delisle *Atlas* Amsterdam: Covens & Mortier, 1739
John Rocque *A plan of the cities of London and Westminster, 1 inch to 200ft* London: John Pine and John Tinney, 1746
G. L. Le Rouge *Atlas portatif des militaires et des voyageurs* Paris, 1756–59
F. W. von Bawr *Campaigns in Westphalia 1756-63,* 1769
John Ellis *Ellis's English Atlas* London: R. Sayer and Carington Bowles, 1766
Emanuel Bowen *The maps and charts to the modern part of the universal history* London: T. Osborne et al, 1766
John Ellis *Ellis's English Atlas* London: R. Sayer, 1773
Samuel Dunn *A new atlas of the mundane system* London: R. Sayer, 1774
Taylor & Skinner *Survey and maps of the roads of North Britain and Scotland* London, 1776
Carrington Bowles *Bowles's new medium English atlas* London, c1785
John Carey *Carey's new and correct English atlas* London, 1787 & 1793
Robert Wilkinson *A general atlas* London, 1794
Bacler Dalbe *Carte générale du théâtre de la guerre en Italie et dans les Alpes* Milan, 1798
Thomas Kitchin *A new universal atlas, 3rd edition* London, 1799
Bacler Dalbe *Carte generale des Royaumes de Naples, Sicilie & Sardaigne* Paris, 1802
John Cary *Cary's new English Atlas* London, 1809
A de Humboldt *Atlas géographique et physique du royaume de la Novelle-Espagne* Paris: G. Dufour, 1812
Matthew Flinders *A voyage to Terra Australis in the years 1801, 1802 and 1803. Atlas* London: G. & W. Nichol, 1814
Carte topographique de l'Egypt...levée pendant l'expedition de l'armée française Paris, 1818
Henry S. Tanner *A new American atlas* Philadelphia: H. S. Tanner, 1823
Ambrose Tardieu *(Atlas of Napoleonic Campaigns 1799-1807)* Paris, 1826
J. Dumont D'Urville *Voyage de la corvette L'Astrolabe...1826-1829 sous le commandement de M. Jules Dumont D'Urville* Atlas *(Planches* Volume 6) Paris: J. Tastu, 1833
Also illustrations from *Planches* Volumes 2, 3, 5
Maps of the Society for the Diffusion of Useful Knowledge London: Chapman & Hall, 1844
The Imperial Atlas of modern geography compiled under the supervision of W. G. Blackie London: Blackie & Son, 1860
J. G. Bartholomew *The English Imperial Atlas and Gazatteer of the World* London: T. Nelson & Sons 1892
The Times Atlas London, 1897

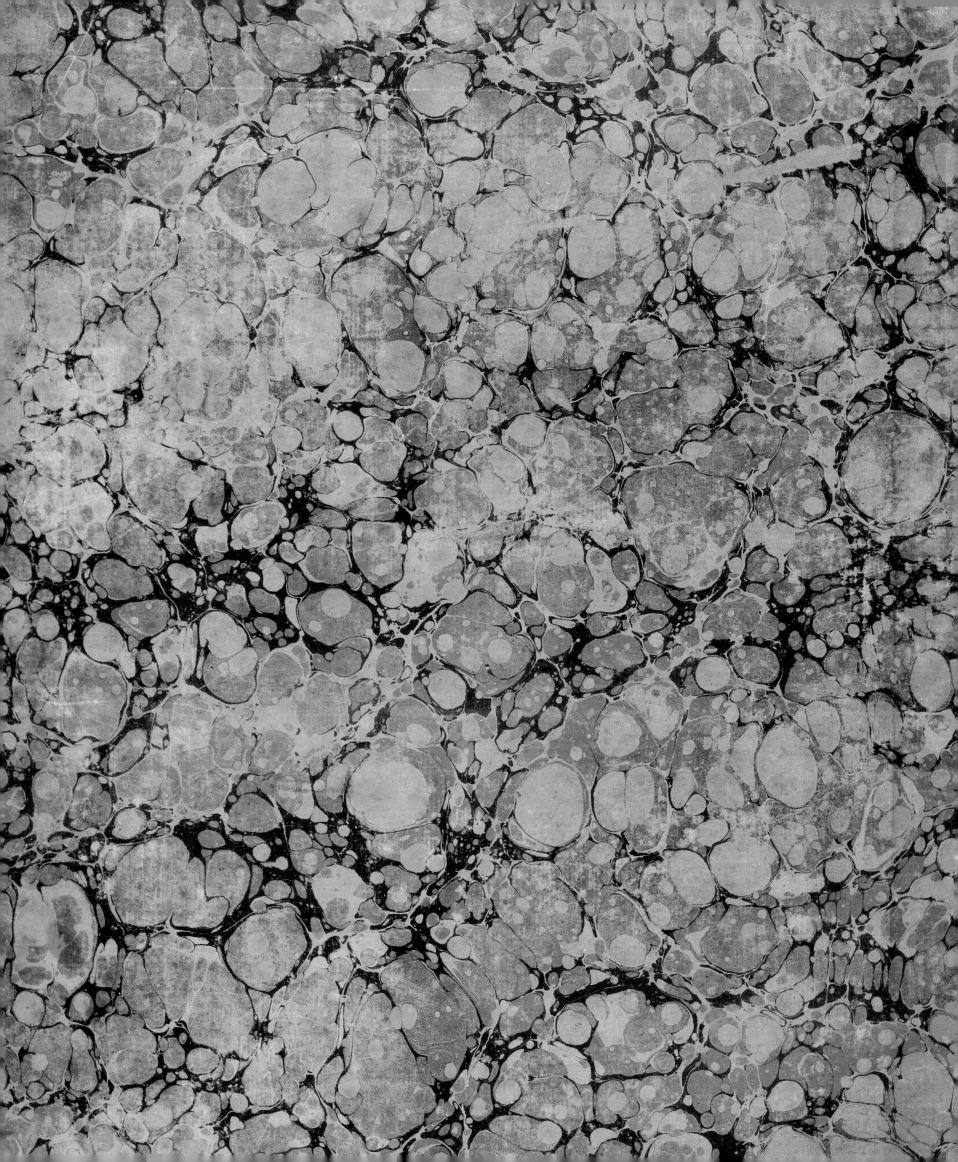